Advance Praise for
Big Tech Tyrants

"Most Americans have no idea of how much information Google, Amazon, and Facebook have gathered on them. *Big Tech Tyrants* is your reference guide into the world of big data surveillance without the jargon. It will help you understand why privacy matters."

—Wayne Allyn Root, Radio and TV Personality

"Floyd Brown and Todd Cefaratti as digital publishers have had a front row seat to watch the changes in technology that have revolutionized: news, politics, and commerce. They will provide you insights into how Big Tech Tyrants use data to manipulate how and what you think. These companies make you dance to their tune."

—Mike Huckabee, Former Governor and Host of *Huckabee*

"I have spent my career in business, politics, and media. These three sectors of our economy have been whipsawed by technological changes. *Big Tech Tyrants* is the most comprehensive exposé I have seen about the dark side of the forces of big data collection and the misuse of it by large tech firms."

—Herman Cain, Former CEO and Media Commentator

"Brown and Cefaratti provide brilliant insight as to how the collection of personal data is daily misused and exploited by the modern-day robber barons of Silicon Valley. Until you read *Big Tech Tyrants*, you won't understand how much data these firms collect and how they use it to manipulate your very reality."

—David Bossie, Deputy Campaign Manager, Trump for President

BIG TECH TYRANTS

How Silicon Valley's Stealth Practices Addict
Teens, Silence Speech, and Steal Your Privacy

FLOYD BROWN
AND TODD CEFARATTI

BOMBARDIER
BOOKS

A BOMBARDIER BOOKS BOOK
An Imprint of Post Hill Press

Big Tech Tyrants:
How Silicon Valley's Stealth Practices Addict Teens, Silence Speech, and Steal Your
Privacy
© 2019 by Floyd Brown and Todd Cefaratti
All Rights Reserved

ISBN: 978-1-64293-290-4
ISBN (eBook): 978-1-64293-291-1

Cover Design by Cody Corcoran

Post Hill Press
New York • Nashville
posthillpress.com

Published in the United States of America

Forewarning

It's early summer, 2020. The presidential campaign is heating up. Challengers to President Donald Trump—from both parties, as well as independents—blast social media boasts in bids to unseat a much-maligned chief executive. Google, Twitter, and Facebook have become the primary battlefields for all the bluster and brag, attack and counterattack. Words and images—jpegs, MP3s, and streaming videos—fill millions of computer screens of all sizes, and iPhones and Androids are bursting their memory limits with the incoming and outgoing tides of political information and imagery.

But all is not as it seems, even to skeptics who have been warning about those seeking to undermine the gold standard of American democracy: the vote. Now, with November 3 in the near future, voters can sense that something is not right with certain candidates. But they don't know why.

One candidate—once known as a friendly, home-loving, hard-working mother of three—somehow now elicits an unsettling impression on Facebook, one of infidelity and deceit.

The YouTube post of a rousing speech given by a beloved Iraq War veteran falls flat on otherwise excited crowds, and one-time supporters begin to have doubts without being able to put their fingers on just why.

Facebook feeds (user-generated content) have changed; they are not quite the same. Long-time friends aren't liking each other's comments as they once did. Newsfeeds are either far more intriguing or far more irritating than before.

Opinion polls begin to describe an electorate that is wandering, confused, uncertain. And the more some candidates work to overcome their sudden reversals in the polls, the less traction they seem to get. But other candidates are doing far better than anyone predicted. Key races are swinging Blue to Red, or Red to Blue. But why?

What is going on in 2020?

Is it possible that some unseen force is adulterating reality, using tools that didn't exist even a decade ago? Tools fashioned of silicon and lines of code, wielded by the darker angels of greed for whom national elections are not about raising up leaders, but about stifling, manipulating, rewriting, repurposing truth and facts, sounds and images, in order to gain financial and intellectual dominion over a troubled and vulnerable land?

Contents

In the Beginning...

All Was Well

Out from a place called Silicon Valley came the modern-day hit parade. Eye-popping new technologies, gadgets that inspired and delighted us, discoveries promising to raise us to levels of prosperity unlike anything the world has ever known.

Busloads of kids who'd once journeyed to Hollywood in search of fortune and fame now found their way to San Jose, Cupertino, and Mountain View, California.

They came to companies named Facebook, Google, Amazon, and Apple. Companies that were bringing the entire planet together in a virtual single town square, an agora of delights. All of mankind's knowledge made instantly accessible. Anything we wanted materialized on our doorstep overnight. Into each of our hands, a supercomputer.

For all of this amazingness, multiple trillions in new wealth was being generated. And all the world wanted to be along for the ride. We all wanted seats on that bus, to bask in the glory of these Mt. Olympus-poised companies, and to offer up sacrifices to their god-like talents. Or at least we were happy to revel in all the Yowza!

But Something Went Wrong

As if we had suddenly been cast out of a bright and promising dawn into a veiled dusk of suspicion, we discovered that one of these companies was running addiction algorithms on its users, stealthily modifying user behavior to soften them up for commercial pitches from advertisers.

Another company came to control what 90 percent of the world sees online and was using that competition-free dominance to punish enemies and push a political agenda on the country.

Then there was the company that refused to pay sales tax, treated employees like dirt, and flattened thousands of competitors, relegating millions of jobs to the dustbin of the unemployed.

And let's not forget the company that aided and abetted terrorism on our native soil by denying federal investigators access to possibly incriminating records, while loyal fans cheered on their new silicon deity.

Something had apparently gone very wrong in that meandering little valley south of San Francisco. We had allowed these tech wizards into the most intimate parts of our lives; we had trusted them. We, the people—from federal, state, and local governments, to corporations, small businesses, and farms, to cities, towns, and hamlets, to mom and pop and even our children—had given them a long leash. But instead of delivering a better world like they'd promised, they reneged on their many public vows and got caught up in the pursuit of "the T Algorithm"—the silicon trophy for being the first company to be worth a trillion dollars.

And in that pursuit, the game changed drastically.

This Is the Story of Golden Boys Turned Tyrants

This true story shows how Facebook and Google, as well as Amazon and Apple, grew so large and oppressive that today they wield unchallenged power over a nation they themselves have impoverished, sitting pretty, high above it all as they systematically dismantle our most cherished freedoms and replace them with…

Well, with what? That's also what this story is about.

STRIKE 1

Addiction Algorithms

An Ingenious Behavior-Shifting Machine

The Meta Tilt, or turning point, in online audience manipulation began in April 2013, on a bright and sunny day nearly shouting with possibility. Especially for the young computer nerd who had found his way from Goldman Sachs through the San Francisco ad tech scene to Facebook, where he was about to pitch Mark Zuckerberg on a new business model the CEO was sure to hate. And this on the eve of the company's upcoming IPO.

Such is the story told by Antonio García Martínez in *Chaos Monkeys*, an autobiography about his time at Facebook and a metamorphic narrative of the valley. It is a story about the approach he and other young techies took to disrupt and dishevel established industries:

- Taxis figuratively driven off the road by Uber
- Hotels replaced with Airbnbs
- Entertainment pumped up with Netflix
- Dating seriously redefined by Tinder
- And on and on…

Martínez called his colleagues "chaos monkeys" for the way they messed with industry after industry in a reckless simian-like fervor. He even wrote, "The question for society is whether it can survive these entrepreneurial chaos monkeys intact, and at what human cost."[1]

1

Facebook is the most successful "thing" ever to happen, measured by audience: 2.2 billion people spend an average of 50 minutes a day on it.[2]

The antecedents of Martínez's pitch that day could be found in the work of B. F. Skinner, the behaviorist from the 1950s who set up a methodical system for prompting caged animals to take a specific action in response to specific stimuli. People were outraged—because it was the first time they had given serious thought to mind control and the immediacy of it in our lives. No longer the realm of science fiction, subconscious behavior modification was becoming a "thing," a real and unsettling thing. Along with subliminal advertising that inserted hidden prompts into the messages all around us (a sexy outline of a woman on top of a can of Coke, swooshing logos, carnivore-like automobile grilles, subtle signals of wealth, power, and action), we discovered that people were trying to mess with our minds.

Yet all of Skinner's behavior hacks and Madison Avenue's trickery were small potatoes compared to today's young techies.

So elementary.

Why try to modify someone's behavior—often against her expressed will—when you can just as easily lay some stealth algorithms on her that she actually enjoys? Now that's tech art—the kind of black magic that Martínez had in mind for Zuckerberg that day.

There were no projectors in the conference room. Zuckerberg likes his presentations printed out and stacked in neat little folders. He's old-fashioned that way. Martínez was cocky (his own words), but also nervous. He knew Facebook's revenue growth had been slowing, its revenue apparently peaking. Advertisers were beginning to question the much-sought-after panacea of "social media marketing" the Valley had been peddling.

Plain fact, it wasn't working for the advertisers or for Facebook.

Insiders Were Questioning if Social Platforms Would Even Survive

It wasn't a question of engagement.

In social parlance, engagement begins with sharing. That is, sharing a story on Facebook is the most "active" thing you can do, like handing a stack of a hundred house-warming invitations to the mail carrier of old. Facebook assumes, reasonably, that when you "share" a story, this kind of story is most important to you. "Commenting" on a story is a little less time-consuming, like jotting a thought on a Post-it Note and slapping it onto a desk, so commenting becomes a little less important. Merely clicking "like" on a story is like a wave to someone across the street—not a big commitment, so even less important in Facebook's rankings of stories you see.

But it all matters.

All of these actions, along with thousands of data points, get crunched in Facebook's machine-learning system to make predictions about what you'll see on your Facebook "News Feed."

All of this is known in the industry as engagement, and Facebook's success in engaging people is why 2.2 billion people spend an average of fifty minutes a day on the site.

It captures our attention today better than any other thing the world has ever known.

And of course, Facebook is only the tip of the social spear. It's well documented that kids are spending an average of ten to twelve hours a day across all digital media. And adults aren't far behind at almost six hours, up from three hours a day in 2009. This includes all the time spent on cellphones, computers, gaming consoles, and streaming devices. Cellphone use alone has mushroomed from a third of an hour a day in 2008 to 3.3 hours currently.[3]

But capturing attention and engaging people is very different from monetizing it all. Facebook was, in its early years, terrible at that. Which is why young Martínez was on the hot seat that bright April day. He needed to produce something to jump-start Facebook, lest the social platform devolve into yet another could-have-been, another flash in the tech pan. He had a couple ideas in his pocket.

One involved using Facebook's Like buttons—what's known as a "social plug-in"—to vacuum up people's browsing behavior on more than half the entire web where Facebook's buttons are positioned. Anywhere you see a little Facebook icon on a webpage, you can be sure Facebook is watching you, seeing your every keystroke, your time on every page, your every action online. But despite seeing everything everyone was doing, the company had no luck monetizing it. They had the richest repository of user data the world had ever seen, but they were not leveraging the data for maximum commercial return. So this first idea of Martínez's was almost a throwaway.

With Facebook watching your every move online, what could go wrong?

Getting to Know You—Really Well

Martínez's next idea was more radical. He proposed that Facebook take all their own data on users and couple it with all the data they could find in the outside world. This would mean tapping into every user's browsing history, online shopping habits, offline purchases in physical stores—essentially vacuuming up a whopping 1,700 points of data that exist on each one of us. Martínez ran through "detailed technical schematics, with walk-throughs of data flows and outside integration points." But he knew all his analysis was being ignored. Zuckerberg was bored by all that; the boy wonder went with his gut.

"You can do this," is pretty much all he said.

Martínez had his marching orders, though it would take another year before Facebook really ramped up on his proposals—the ones that would cause the biggest firestorm the fledgling online world had ever seen.

That firestorm would come first from a product Facebook called "Custom Audiences" (and is similar to a product Google has). Custom Audiences would have so much of your personal data, they could identify you in a police lineup 90 percent of the time. Nobody in the advertising business had ever accomplished anything even close (though not for lack of trying).

To a marketer, this was the Holy Grail. Martínez put it something like this: If you show up bleary-eyed at eleven p.m. at your local Target, they know if you are looking for tampons or a six-pack of Bud Light.

Piling onto the firestorm to come was the second product, "Facebook Exchange." This was algorithms taking over digital media, creating a New York Stock Exchange of eyeballs where our every human wish, hope, want, need, or dream was commoditized and traded tens of billions of times a day—hundreds of thousands of times a second—all for money.

And lots of money. Because with Custom Audiences slaved to Facebook Exchange, any advertiser could dynamically fine-tune any ad for any user on any device at any time of day. It was like being in the users' minds, knowing them better than they knew themselves.

Yes, just like that.

And when this ad-serving juggernaut was thrown against Facebook's News Feeds on the mobile app, the lid blew off Facebook's money-making potential. Not necessarily a bad thing, except the lid also blew off Facebook's ability to control the mischief that could play out on its platform.

Such are the fruits of unintended consequences: some sweet, some sour, some rotten.

This was an important time, historically. Our culture was changing. Cellphones were going mainstream. Their small keypads were becoming the touchstones of a new religion, an ever-present chiclet rosary for a growing generation of willing, eager believers. These devices were suddenly everywhere. With everyone carrying a cellphone all the time; with teens even sleeping with theirs; with a majority of girls telling researchers they would give up their boyfriend before they would relinquish their cellphone, Facebook had the perfect behavior modification platform designed to detect, examine, react, and provide feedback on the most mundane, or the most intimate, of human activities.

Users could now be constantly tracked and measured and unknowingly given cues and prompts on a steady drip, all custom-tailored. Users could be hypnotized little by little by technicians they'd never see, for purposes they may or may not approve of. They could be reduced to prompt-and-respond lab animals instantly recognizable by B.F. Skinner himself.

Which is exactly what happened.

Twitter sells conflict, Instagram sells envy, Facebook sells you.

—WALTER KIRN[4]

And with that, the entire online experience went south. Not just for the users, but for concerned parents and anyone who cares about our culture. Because this was not simply a matter of adults deciding to open up their lives online in exchange for a networking experience. People were now giving up something that has defined us as social creatures for millennia: our ability to self-set the parameters of our basic privacy. And by "basic privacy" we mean the expectation that our personal life is ours and not something to be constantly surveilled, pricked, and sucked into an algorithmic syringe expertly operated by corporations that make money by manipulating our behavior. These techniques reduce us to a handful of pixels to be ripped apart online, bought and sold and parsed twelve ways to Sunday, sometimes causing us no real harm, but other times opening us up to all kinds of liabilities and bad, even dangerous experiences.

This new kind of data mining is little different than injecting Botox: the results can be beautiful…or they can be hideous.

When Facebook takes all the data points they have on us, and packages them for outsiders, the process can lead to any number of outcomes. But most often, one of two things happen:

1. We get messages targeted very closely to our interests from merchants we care about. We get them so often it can become annoying, but there are worse problems to have.

2. Aspects of our lives that we'd prefer to leave private, that we'd assumed are private, are made public, causing us all manner of harm.

When you think about all the things Facebook knows about you, you begin to get a finer appreciation of the real cost of lost privacy.

Here's What Facebook Knows About You

Here's what the algorithms gorging on data about you are piecing together:

- What other sites are you visiting on the web?
- What kinds of links do you click on regularly?
- When do you switch from public to private mode, and for how long?
- Which videos do you watch partially, which all the way through?
- How quickly do you move from one page to the next, one site to the next?
- Where are you physically when you do these things?
- Who are you connecting with online, and then in person?
- What facial expressions do you make while online (yes, this involves your computer's built-in camera)?[5]
- How does your skin tone change in different situations?
- What were you doing just before you decided to buy something or not?
- What are your political affiliations, and how active are you in politics?
- Do you vote regularly, and if not, why not?[6]
- Where do you live; where are you most likely to go next?
- Where did you go to school; do you keep in touch with old classmates?
- Are you single, married, about to change that?
- Are you questioning who you are, your sexuality, your gender?
- When's your birthday and the birthdays of all your friends?
- What's your current job; are you happy at it or looking to leave?
- What are the birthdates of your children, and how do you celebrate them?
- What did you spend money on in the last minute, day, week, year?

- When are you likely to die?
- How do all of these data points on you cross-reference with everyone you know?

Facebook says this data collection is essential to their mission—to create a more relevant set of connections for you, and to help advertisers and developers serve you better-targeted ads and apps. Facebook has said, "When we ask people about our ads, one of the top things they tell us is that they want to see ads that are more relevant to their interests."[7]

But they said that in 2014. Before young Martínez convinced Zuckerberg to dial up the addiction algorithm. Before Facebook went looking for "partners"—thousands, even millions, of them. All with an agenda. So even if you can trust Facebook to treat you right, you don't know about all its partners who earn that special distinction by writing Facebook a check.

Along with advertisers, third-party app developers were granted admission onto Facebook's platform. That created a two-way street where the developers had access to Facebook users and the games, quizzes, and dating apps they developed and uploaded.

The Potential for Mischief Was Always Known

What outsiders can do, and have done, with the Facebook data feed is ominous. Some sociologists at the University of North Carolina ran a study to show how much damage could be done on Facebook. They studied only Facebook Likes and didn't venture deeper into the platform's data treasure trove. Just using Likes, they found they could predict, with 80 to 90 percent accuracy, the "latent" traits of 58,000 Facebook users who volunteered for the study. These traits included religious and political views, sexual orientation, ethnicity, personality, intelligence, happiness, use of addictive substances, and closeness to parents along with the usual easy markers like age, weight, and gender.[8]

Fancy Ad Targeting Was Only the Beginning

These algorithms are constantly crunching on you, trying to figure you out, but it's actually not accurate to say that they *understand* you. It's more about there being *power in numbers*. The larger, the more powerful.

For example, say the algorithms that are constantly crunching your data learn that people who like the same movies, food, and sports as you are not fans of a political candidate when that candidate is described in a news story using the font Verdana instead of Helvetica. So if the guys at the social platform don't like that candidate, the font you see in your News Feed can be changed to Verdana—because odds are you will like that candidate less, as well, if you see the story in Verdana font.

Seems like a little thing, right?

So far it is, but hang on.

If this same slight tweaking is done over and over again on a multitude of factors by algorithms that never sleep, then your behavior can be subtly shifted and you are totally unaware. Nobody may know why a different font has this effect, and it doesn't matter why it does. Only that it does. Because statistics are reliable. And so the candidate can be presented to you in the right font, right colored background, right framing technique, so that you form an opinion about that candidate that's in line with the unseen manipulators' goals.

This may sound like a stretch to you. But we know people are susceptible to the slightest shifts in their environment. We are all impacted by the shifts—even if we don't know it. You might even find this idea insulting, because we are suggesting that you're being turned, little shift by little shift, into a well-trained dog, or lab rat. Being remote-controlled by unseen manipulators, or, if you prefer, puppeteers.

But that's exactly what's happening. And they've normalized it; that is, there is now a business model attached to this.

A model of surveilling people and manipulating them in unseen ways on an ongoing basis. A model that essentially turns social media platforms into crime scenes, holding guns to users' heads, rifling through their intellectual and emotional pockets, and escaping through a well-designed maze. You have to consent to the platform's terms because there are no real alternatives.

It's an emotionally manipulative model that manipulates your feelings. This becomes especially dangerous when you consider that, increasingly, more people—especially young people—are actually deciding what is true based on their feelings, and they think feelings are more important than facts.

This is activity we used to consider unethical, inhumane even. But no longer, as people like Zuckerberg are re-engineering who we are.

People have really gotten comfortable not only sharing more information and different kinds, but more openly and with more people. That social norm is something that has evolved over time.

—MARK ZUCKERBERG, Facebook CEO

Getting to the Core of an Eye-Watering Problem

In order to fully understand and address the massive problem the Big Tech Tyrants have unleashed, it's critical to define the problem very clearly. Doing this is not easy. It's kind of like peeling back the layers of an onion—your eyes water a little more with each layer, but you need to get to the core of it.

Is the problem, as some say, that there are now literally billions of strangers crammed into online environments that are (technically) stone-cold and lacking the genuine community and empathy that's found in face-to-face interactions, and so the worst in the human character is thrown onto the online display? Or is that only part of it, because these online conversations we have can often bring out the best in us, especially in times of crisis?

Is the problem then, as others say, all the controlling strings in our lives are now concentrated in the hands of just a few puppeteer companies? Or again, does that only partially explain it, since big companies are not necessarily bad companies, just as little companies are not necessarily good companies?

Is the problem, as still others say, that we now all carry pretty much the same smartphone with us and these little gadgets are optimized for mass behavior-shifting? Or is that only part of it, because surely our smartphones, iPads, Kindles, and other devices are good for other uses, as well.

Or could the problem actually be a new business model being worked out in Silicon Valley that turns people into products that others pay to manipulate?

So which of these concerns best explains the state we're in? Too many people crammed into an impersonal medium? A few big companies calling all the shots? Our having devices in our hands for 24/7 behavior shifting?

We believe that all of these things help explain the state of online affairs, but that it's a corrupt and even evil business model that's at the core of this eye-watering onion.

In the olden days, say five years ago, advertisers would measure their results by how sales or awareness, or another objective, increased after a given ad ran. But with the new business model, the platforms (Facebook and Google primarily) count up the number of "stealth shifts" they've delivered. The more, the better, presumably. That indicates that the platform is working the way it has been engineered—explicitly from the perspective of that platform's operator.

It's not sensationalizing to call it evil at the core because the "stealth shifts" that pay out best for Facebook and its partners are the negative ones.

The sadder, madder, more infuriated you become, the more time you spend on Facebook, and the more money they make.

The more they dial up the hottest buttons of discontent and discord, the more money they make.

The more harm they do to the social and political fabric, the more money they make.

Through machine-learning algorithms—what's known as narrow artificial intelligence—they run through countless terabytes of data in an effort to impact our emotions and nudge us toward an outcome they desire. And it's all done independently of our choosing, or even our awareness. Which is why, quite clearly, the opportunities for grand mischief scale up exponentially.

It's a degrading, depressing business model.

The platform's objective is to mix in just enough positive emotions—along with an overwhelming wave of negative emotions—because that's what keeps users using. If folks were to grow happier or more pleased after their time online, they might take a breather from their obsession with social media and, God forbid, go out and smell the flowers or ride a bike or engage in some other revenue-crimping activity.

Better, then, to keep them perched on the edge of their seat worrying about their social media ranking numbers or whether the world is coming to an end or the terribleness of their political opponents. All things guaranteed to keep them engaging—clicking, commenting, poking, swiping or whatever inventive new "engagement device" the engineers dream up next.

> **An aside from the gambling underworld:**
> **These online exploitation algorithms were not pioneered by**
> **Facebook and Google, but by the gambling sites to "hook suckers."**
> **These gambling sites admit to being angry at the rip-off,**
> **but ecstatic at how effective social media has become**
> **at identifying easy marks for them.**[9]

In this sense, these social platform operators are like the corner dealer, doling out nickel bags of weed for free to kids, getting them good and glassy-eyed before handing them off to a "partner" who has something harder to sell them.

This is Facebook's business model—and Google's as well. It's the culmination of the young quant Martinez's labor as a self-described "chaos monkey" in Silicon Valley. Sowing so much chaos that no simple solution will un-mess the mess.

That doesn't mean it can't be done. Or that we mustn't find a way.

Other companies now reach out to customers and prospects using personalized messaging, and that can be a good thing. When eBay recommends an item we may want to purchase, or when Netflix's algorithm offers up a movie choice for us, they are providing a value-adding service. They are helping us *make a decision*. They are not being paid by unseen partners to "stealth shift" our behavior independently from, and invisible to, our very reason for being on the site.

These algorithmic shifting technologies are far more troubling than dark ads (seen only by the person they are targeted to; nobody else knows it happened) from Russian, Chinese, North Korean, or Iranian propagandists.

Fact is, governments have been meddling in other countries' elections at least since 1954, when the US tried to influence the outcome in Iran. Psychological operations (PSYOPS) are a valuable battlefield tool for military and intelligence personnel to employ as a technique to affect the "hearts and minds" of enemy combatants and civilian populations. Changing minds to improve political, social, or philosophical outcomes has historical models reaching back thousands of years. Only the technology has changed.

The outrage you hear about Russian trolls in Washington is just a page from the political playbook of an election loser or party out of power. Had the last election gone differently, the other party would be howling about it. It's the Kabuki theater of Washington, just another curtain call. If the Russians were acting as electoral terrorists, as some have charged, then their crimes are misdemeanors compared with the daily business model of Facebook.

That's something that's arguably criminal, as we'll see.

> **Facebook is better named Facehook.**
> **It's got its barbs in 2.2 billion mouths.**

Shifting Us Into Separate Realities

When each of us sees a different world that Facebook serves up in our unique News Feeds, we can lose touch with the bigger reality (68 percent of Americans get news on social media).[10] The social cues that used to be built into our daily lives fade away. Our perception of true reality beyond the platform suffers. And then turns dangerous.

Take the case of the North Carolina man who bought into the rumor generated during the 2016 election that Hillary Clinton was running a child sex ring out of the basement of a pizza shop in Washington, D.C., so he went in guns blazing. Was he nuts? Was he an idiot? Was he whipped up by a false belief spread so easily online?

Yes. He may have felt like Truman Burbank in the 1998 movie, *The Truman Show,* where an entire reality was created around one man so people could be entertained by him.

Of course you can go back in time, to any period, and find similar acts of crazy from people who appeared "cut off" from reality. The Salem witch trials and the Inquisition come to mind as they were about persecuting people without any evidence of wrongdoing. The trials were based more on emotion than facts. But these things happen often now. Do the Justice Brett Kavanaugh hearings ring a bell? Brett Kavanaugh was persecuted for political reasons and not factual reasons. The persecutor's "evidence" was based on hearsay and emotions and not factual evidence. Bottom line is the persecutors didn't like the political beliefs of the one being persecuted even though they lost the election and the president had every right to appoint justices who he wanted. They happen with such maddening regularly that it appears many people are no longer living in the same world as the rest of us.

Think of it this way. A thought experiment, if you will. What would happen if the Department of Motor Vehicles in every county and every state gave a different driving test to each applicant? Some people would be told the city speed limit is twenty-five, others forty-five miles per hour. Some would learn that traffic turning left has the right of way. And so on down the line— everyone getting different rules. How would that turn out?

We know that accidents, countless quarrels, and madcap (or deadly) confusion would reign.

But as bizarre as this sounds, it's pretty much what we've done in the new online marketplace…all mediated by the social platforms. It's the new normal. All across this new agora, there is a rattling of assessments being made about each of us, impacting each of us, mostly unknown to each of us.

How many friends do we have?

How many followers?

Are we hot? Or not?

How many affinity points have we earned?

How many virtual gold stars or pink badges or some other invented reward have we gotten for actions we've taken, content we've generated, polls we've taken to build up the social platforms?

Social platforms call these questions intermediate-layer interpretations, and they are output by the algorithms. They are used to optimize ad delivery—to learn which ads will have which effect on you. The more questions they can answer about you, the better they can target ads to you. So the algorithms are constantly at work trying to decide:

- *Which news stories hold you longest?*
- *Which beached whale photo holds you?*
- *Which calico cat?*
- *Which refugee child?*
- *Which family members?*
- *Which former flames?*
- *Which new gadgets?*

Which, which, which…in an eternal whirl of intelligent algorithms giving you the precise mix of content to keep you clicking, tapping, and scrolling longer. And all coming to an algorithmic crescendo—your ranking in the online world, your value, your worth to the social platforms.

This "which, which, which" is the guts of the addiction algorithm, tapping a neurological process we barely understand. All we know is that the neurotransmitter dopamine is the chemical that helps control our brain's reward and pleasure centers and gets really involved in changing our behavior in response to getting rewards. So each little "which" becomes a little dopamine hit that we want to keep striving for—modern-day rats on the treadmill.

But the platforms' addictiveness goes well beyond these positive dopamine hits. It's not the whole story. Because the platforms use punishment and negative reinforcement mercilessly.[11]

Sorting Us Into Depressing Isolation Boxes

In countless TED Talks, the Tech Tyrants have waxed eloquent about making our world a happier, more connected place—and yet the science says otherwise. In our real-world associations, we've grown more isolated than ever, or at least most of us have. If you'd like to see the academics' take on this, we've recommended some articles.

"Social Media Use and Perceived Social Isolation Among Young Adults in the U.S." Lead editor: Brian A. Primack.
Published: *American Journal of Preventive Medicine*[12]

"Facebook's Emotional Consequences: Why Facebook Causes a Decrease in Mood and Why People Still Use It." Editors: Christina Sagioglou, Tobias Greitemeyer.
Published: *Science Direct*[13]

"Facebook Use Predicts Declines in Subjective Well-Being in Young Adults." Editor: Cédric Sueur.
Published: *Institut Pluridisciplinaire Hubert Curien*[14]

"Association of Facebook Use With Compromised Well-Being: A Longitudinal Study." Editors: Holly B. Shakya, Nicholas A. Christakis.
Published: *American Journal of Epidemiology*[15]

Odd as it may sound, Facebook's researchers have actually been caught bragging about this thing they call "isolation boxes" or "filter bubbles." Yes, they actually tooted their horns in public over their ability to isolate their users and inflict torments that anger or sadden the users, without them realizing why.[16]

Why would they brag about tormenting people...and so publicly? You'd think it would be bad for the brand.

Or maybe they knew exactly what they were doing: periscoping their product offering. After all, as we've seen, Facebook has users (the product) and partners (the revenue source). Facebook wants those partners to know what they are capable of.

These large corporations (and governments and political campaigns) now have new tools and stealth methods to quietly model our personality, our vulnerabilities, identify our networks, and effectively nudge and shape our ideas, desires and dreams.

—ZEYNEP TUFEKCI, Sociologist at the University of North Carolina[17]

Changing What Happens to Us in Real Life

For those who might dismiss these concerns with a blithe, "I'm an open book—no secrets—don't care what they know about me" attitude, let's look closer at how much these behavior-shifting algorithms impact you in real life.

They are responsible for the news you see on Facebook and Google, as many people do, but you can always get your news elsewhere, right?

Then they're responsible for the type of people you get introduced to as potential mates or business partners, and also the kinds of products you have set before you. Again, you know you have options.

Not so with the judgments algorithmically made about you on the social platform. Here you'll find decisions being made about you that can open up or close off:

- *Your next job position—from even getting an interview.*[18]
- *Your next car or home loan—from closing for you.*[19]
- *Your next auto insurance claim—from going your way.*[20]
- *Your next overseas trip—from being allowed into certain countries.*[21]
- *Your children's future schools—from being allowed to attend.*[22]

These are all examples of how third parties are relying on the data assessments received from Facebook or Google instead of doing their own due diligence work (like they used to). They're tapping your social feed, and so, yes, the contents of that feed have become very important.

There are still more examples. And some are disturbing. It's known that the government tracks the attendance of people at political rallies. They do it from data handed over by Facebook, Twitter, and Instagram.[23]

Your postings can even get you killed. This is an extreme case, but telling: Back in 2007, some enemy soldiers took geotagged photos of Apache helicopters arriving in Iraq; insurgents followed the online posts to the helicopters and took them out in a mortar attack. US assets have been targeted with the help of social media for more than a decade. It continues today all around the world.[24]

Outright Weaponizing the Social Platforms

Earlier we touched on how these AI (artificial intelligence)-driven algorithms could make a bunch of little shifts to affect a major shift.

The social platform designers have found that changing the font used in a photo caption or the color of the border around the photo could make a given percentage of people trust a candidate less or more. Just font and color, that's all. But that's making a difference that nobody outside of the platform (or, perhaps, nobody inside the platform, since it's machine-driven) would even know about. Once the order was given, "Shine this candidate in as bad a light as possible, but leave no tracks," the algorithm would take those instructions and go to town, spinning through every possible permutation of trillions of optional data points to find the ones throwing the most shade on a political opponent.

Why does the font or the color make a difference? Maybe the font was associated with a widely seen story of horror that day, or some other reason. It's well-documented the effect different colors have on people's emotions, i.e., "true blue." A 2014 study examining the effects of color on moods cited various researches that "the color that surrounds us in our daily lives has a profound effect on our mood and on our behavior."[25]Some colors induce positive emotions (respondents' in the above survey's favorite color was blue; blue is calming and associated with logical thought), while other colors are negative or stimulating (red is associated with aggression, "fight or flight," and power). It's all in the numbers—the terabytes of data—which don't lie.

Each little shift in behavior that's triggered adds over time to a big shift, like the effect of a lowly copper penny that's doubled every day and, after a month, is worth more than five million dollars! This is how behavior shifts rip apart the bonds of society and lead to a widespread tribalism marked by self-interested behavior.

It's built right into the shifting algorithm, right into the code, seizing upon any latent prejudice buried in our heads—deep for some, shallow for others—and then scratching at those little neural memories, creating an itch, and surfacing bad behavior.

The Four "Dark Patterns" of Manipulation

There are four dark patterns, summarized by Consumer Watchdog, that Facebook uses to trick, cajole, and coerce consumers into unknowingly revealing valuable data about themselves:

1. Privacy-intrusive default settings: Facebook and Google set the least privacy-friendly choice as the default option. This is a problem, given that research has shown that users rarely change pre-selected settings.

2. Illusion of choice: Users do not have an option to opt out of disclosing their data entirely. Disclosure of personal information is a condition for using the service. Yet Facebook and Google create the illusion of choice by providing "the feeling of control" when the choice is actually very limited.

3. Hiding privacy-friendly choices: Privacy-friendly choices require significantly more clicks to reach and are often hidden away.

4. Deceptive design choices: Disclosure of personal data is presented as beneficial to users, often in combination with threats of lost functionality of services if users decline.

As a result of these dark patterns, Facebook users have been shuttled onto isolated tracks and users can spend so long on those tracks, they lose any sense of peripheral vision. We can very easily begin to develop a somewhat distorted worldview. And we don't know how it's being done to others around us, or how their views are being distorted. All of this is opaque to us all.

They have our attention. They cleaved us off from the herd. We no longer know what others are thinking or doing. Sure, there is some overlap, and there are other ways to get news and views. But more and more, we no longer have shared experiences, for those with whom we once shared common rails are now being manipulated and diverted to their own separate tracks. Where once we, as a common-bond society, saw the same signal lights of progress or danger, now we see individual signals custom-designed for journeys we didn't even know we were going to take—sometimes toward progress; more often, lately, toward danger.

This is the tilt in our new algorithmically designed town square: The world each of us sees (or that we think we see) is visible only to us. It's not visible to others, and how can that *not* lead to misunderstandings, political handwringing, and a fracturing of society's center, itself no longer capable of holding?

Content is chosen for us, ads are customized to us, and we don't know how or why. What we see is very different from what our friends see, what our neighbors see. And none of us has any way to know what others are seeing. Each social platform feed is unique; we become isolated in what are known as "filter bubbles."

Over time we know less and less about each other because we've lost a portion of our shared interests and experiences. Our once-mutual sense of empathy begins to diverge and we gradually—or not so gradually, it has turned out—lose the ability to understand one other.

Astrophysicists tell us that the universe is expanding and accelerating, and that all the galaxies we can now see are rushing away from each other, eventually to disappear over the visible horizon of space-time. One day, under this scenario, the night sky will be empty, all other worlds having slipped beyond our view. So, too, have we begun to slip away from each other as the greed-motivated moving hands of technology drive us apart—soon to be alone, by design.

Empathy is the fuel of civil society. Without it, we are left with dry rules and competitions for power. We are left feeling as if life has become a near-Hobbesian reality—nasty, brutish and uncivilized, though not short. If anything, the growing coarseness of the online experience can feel as if it will never end.

Of course, we can fight the filter. We can seek out the kinds of content that other people are probably seeing. We can keep up with political views we may disagree with—and no longer see in our own feed. We can seek out people in real life! But it's an uphill battle. It takes a kind of rigor that few of us have outside our normal routines.

We're fighting fine-tuned algorithms that know what they're doing. Do we have a chance?

The practiced ease with which this new business model can shift emotions in addictive and manipulative ways just screams "trouble ahead, all the signals are red," as we've seen. Weaponized information can sway elections, help hate groups recruit, and give society's refuse the tools they need to lay minefields of social discord.

Outside of China, this is not a popular business plan. Only Facebook and Google truly rely on it for almost all of their revenues. Other Big Tech Firms like Apple, Amazon, Twitter, Netflix, and Uber slip into this behavior-shifting mode occasionally as well, because it has been normalized. But they are not dependent on it.

In their simplest models, they stealthily take data from you and make money off it. Their wealth is made entirely of the data you gave them. Call us old-fashioned, but we believe a company should get rich by making things people want, not by making people feel less than adequate by showing them what they don't have.

Capitalism was never meant to be a zero-sum game.

Bumming out your users for profit is a nasty—some would say evil— game. And so clearly the biggest names in the Valley are nasty and evil.

"Let us spy on you and secretly manipulate you in return for free stuff" is not a business model any nation should be proud of. We can do better, friends.

You can't make a society wealthy by making it crazy.[26]

—JARON LANIER, Scientist and author of the *Ten Arguments for Deleting Your Social Media*

Just because this terrible business model has been normalized doesn't mean it must endure. After all, the model is still in its diapers stage—started by Google and perfected by Facebook beginning in 2014. Just as a toddler teeters hazardously between falling and walking (and grabbing the dining room tablecloth) until their motor skills kick in, this business model will, we hope, mature into a solid citizen contributing to society.

Maybe a platform's users can be convinced to pay to use the platform when, unthinking or unknowing, they practically consider it an entitlement. Maybe another company will come along with a healthier model and shoulder aside these companies who now own the world.

And maybe pigs will take to the skies in flight.

These platforms are not going to change unless they are made to change. In later sections we'll talk more about how that can happen, and how America can win in the deal.

Facebook spokespeople say their company "complies with the law, follows recommendations from privacy and design experts, [and helps] people understand how the technology works and their choices."[27]

Google's people say, "We've updated our privacy settings 'over many years to ensure people can easily understand, and use, the array of tools available to them.'"

But the Electronic Privacy Information Center and other advocacy groups say in a letter to the Federal Trade Commission (FTC), "We urge you to investigate the misleading and manipulative tactics of the dominant digital platforms in the United States, which steer users to 'consent' to privacy-invasive default settings."[28]

The Norway Council of Consumers added:

> Facebook "gives the user an impression of control over use of third-party data to show ads, while it turns out that the control is much more limited than it initially appears," and that Google's privacy dashboard "turns out to be difficult to navigate, more resembling a maze than a tool for user control."[29]

Now the FTC is looking at these tactics. John Simpson of Consumer Watchdog, along with other privacy groups, is questioning:

> Did Facebook's "tactics constitute unfair and deceptive trade practices under Section 5 of the FTC Act" and so if "the FTC needs to take a stand

against Facebook and Google for deceiving the American people, as well as Europeans, into giving up their privacy."[30]

We join with these privacy groups in urging FTC action. But that alone is not likely to make any significant difference in how these platforms harvest our information and sell out our privacy like we're little more than green lines of code from *The Matrix*.

That kind of solution will have to come from a higher level.

> **Facebook is under such scrutiny by regulators in the US and Europe as we write, they are constantly changing the screens you see, what's known as the "user interface." So these screens may have been scrubbed and sanitized to appear less controlling, and more caring. But the business model has not changed— that's what matters.**

Deciding What You'll See (or Not See) in This World

Few who would pose as modern gods succeed. Certainly not on the level of Google of Mountain View—our modern Mt. Olympus. From its rarefied heights, Google has become humanity's oracle, its source of wisdom and knowledge of all that exists in the universe.

It knows just about everything about us—down to our deepest secrets. It tells us where we are and where we need to go. It answers our every question, from the "time of day" to "how to build a time machine." If we are to judge Google by the questions we ask of it, our willingness to believe in it, and our surety in following its guidance 3.5 billion times a day, there is no institution from the oldest church down to the local PTA board that we hold in higher esteem and implicit trust.

A subsidiary of Alphabet Inc., in 2017 Google earned $109.7 billion in profits, which they've used to wipe out one traditional brand and media outlet after another. Nobody has been able to compete with the Google god. How do you compete with a machine that can deliver pretty much any answer—no matter how difficult the query—in just .000003 seconds?

But are we getting pretty much what we're looking for or, unknown to us, are we actually getting what Google wants us to see—based on Google's own business objectives, not based on what we asked for?

Few have questioned how Google's results are obtained and packaged for us. It's hard to argue with the amazing convenience of it. That convenience is one of the great accomplishments of our time. It is why Google has become a modern god in our lives; why we take it on faith that this modern god cares for us. But what is this faith based on?

What are these wizard gods doing behind the curtain?

Google Is Profiling Us for Fun and Profit

For starters, Google is creating a profile of you—whether you want them to or not. It's called a "shadow profile," and like the dark companion it suggests, it goes on in the shadows of your daily life, with or without your consent.

Google allows everyone, whether they have a Google account or not, to opt out of its ad targeting. Yet, like Facebook, it continues to gather your data.[31]

This matters because of how these shadow profiles can be used.

This matters because of a tool called Google Analytics, which is used by more than half of the Fortune 500 companies and an estimated 50 million merchant websites. It is used by so many precisely because it's so good at what it does. It tracks everything on the internet…right down to *you*. Whether you are logged in or not. Whether you have opted in to tracking or not.

Like it or not, ask for it or not, you are being tracked. When Butch and Sundance were being hunted down by paid trackers, they frequently took note of their dogged pursuers' intensity with the remark, "Who are those guys?" Today, you don't even have the advantage of seeing your pursuers, but they most assuredly are on your trail.

This tracking continues onto your mobile devices. There again, Google is always listening, always watching, always recording, and doing it almost everywhere you go.

Because it owns the operating system, Google is able to constantly harvest data through the world's two billion Android mobile devices. Android users of Gmail will keep getting asked to enable access to the device's camera and microphone until they say yes. Android users will also be asked by the Google Maps app to turn on "Location Services." Like with the free Gmail accounts, the free Maps is a very useful tool—which is why so many of us use it. But these services are more than a way to serve ads to you. They are a way of knowing what you're doing all the time.

Google can cross-track devices to almost instantly identify just about everyone in the world no matter where they are or which device they're using.

They can determine who—and where—we really are, whether or not we reveal ourselves voluntarily.

Meanwhile, the billion-plus people who have Google accounts are tracked in even more ways. Google uses our browsing and search history, apps we've installed, demographics such as age and gender and, from its own analytics and other sources, where we've shopped. Google says it doesn't use information from "sensitive categories" such as race, religion, sexual orientation, or health. But that becomes a semantic distinction.

That's because Google's data-harvesting capabilities actually extend to the thousands of data brokers in the US who have up to 1,700 data points on each of us and collectively know everything about us that we might prefer they didn't—such as whether we are pregnant, about to get divorced, or trying to lose weight.[32] Google works with these brokers directly, selling them ad-targeting information.

Google says it vets these brokers to prevent targeting abuses. This vetting process must begin and end with their ability to pay their bill net 30. Because hundreds of these outside partners buy up all the Google data, they can build precisely the kinds of "sensitive information" files on people that can lead to privacy abuses.

That's the business they're in. Selling this information to insurers, bankers, competitors, employers, political groups, and anyone else who might be interested.

Google also delivers to data brokers what's known as "Lookalike Audiences"—profiles of people who are similar to the people a broker may be targeting for an ad campaign. This means you can be targeted with ads even if you have never shown an interest in a product.

This is a fairly tame practice compared to a secret effort Google ran to open up your most private conversations to anyone with a checkbook.

As we learned in a *Wall Street Journal* investigation, hundreds of outside developers have been allowed to scan through more than one billion users' Gmail accounts using their own third-party apps.[33] In some cases, these developers could look right into the Gmail inboxes, into the messages—even as they were being composed by unwitting users—and peruse away to their heart's content.

Sensitive information and all.

Two marketing companies, for example—Edison Software and Return Path—were given the keys to Gmail and allowed to read thousands of emails.

Their intentions may have been entirely honorable and solely in the interests of improving their own products. But what about those who weren't?

When the *Journal* broke this story—and only then—Google was quick to announce that they would stop letting marketers scan users' email for any reason. Google buckled under just days after having insisted publicly that they don't open up Gmail to developers for any reason. As per Google's press release:

> "The practice of automatic processing has caused some to speculate mistakenly that Google 'reads' your emails…To be absolutely clear: no one at Google reads your Gmail, except in very specific cases where you ask us to and give consent, or where we need to for security purposes, such as investigating a bug or abuse."[34]

Nobody should have been surprised that Google had been lying. But more than that, for us anyway, we got to wondering what Google meant when it referred to this "practice of automatic processing." What could that mean in this context?

Is it a made-up phrase, intended to further obfuscate, or to somehow justify themselves? And whatever it means, it would suggest that there are no humans involved, it being automatic and all. But the very act of letting human developers read human users' emails is the opposite of automatic processing; it's manual processing, by definition. So was Google wrapping a misdirection inside a fabrication to hide a lie?

Perhaps the best answer comes from all the money Google, along with Facebook and Amazon, have been spending to lobby US politicians. Their lobbying has most directly been targeted at liberal politicians who publicly decry abusive, monopolistic corporations while privately taking record donations from those same corporations.

Halfway through 2018, public disclosures showed that Google had spent $5.83 million, Facebook $3.67 million, and Amazon $3.47 million in buying political cover for themselves—and the bulk has gone to Democrats.[35]

Google knows the game it is playing and which party is more likely to provide cover.

The company actively opposed the California Consumer Privacy Act as it was headed to the voters. The act was intended to grant consumers three basic protections:

"The right to tell a business not to share or sell your personal information, the right to know where and to whom your data is being sold or shared, and the right to know that your service providers are protecting your information."[36]

Into 2018, even after Facebook dropped its opposition to the act when it became clear the bellwether California voters cared about it, Google persisted. The act itself was later superseded by legislation. But yes, Google well knew the game it was playing.

That's why they hire guys like David Goodfriend, a Washington lawyer and former deputy staff secretary to President Bill Clinton.[37] Goodfriend brags that he "fights mergers for a living" and that "strong antitrust enforcement is in my progressive DNA." So when he was first asked to represent America's truest monopoly, he demurred, saying it ran against his principles. Yet he did manage to say yes to Google. Why?

Goodfriend says it's because of "the alt-right's rise," and he knows clearly his dog whistles. He knows that ascribing his actions on the alt-right threat gives him a pass.

It sends a message out to all good progressives. The alt-right is attacking Google; therefore, I must defend Google. Goodfriend insists he "saw a real danger in the alt-right's thinly veiled use of antitrust as a way to score political points" and an "alt-right smokescreen" in an op-ed he wrote. He positioned Google as "freethinking, open-minded people with power who are willing to resist." Resist what, you might wonder? He didn't say.

It no doubt meant resisting the alt-right, which, in his mind, includes everyone who doesn't vote the Democratic Party line. So half the country, or thereabouts, is left out of the Google vision for America as defined by their chief lobbyist, David Goodfriend.

The liberal lawyer also tried to demean conservative criticism of Google by asking: Why are conservatives upset about Google's antitrust activities and not the Sinclair Broadcast Group—a conservative organization, to be sure, and one that owns a lot of TV stations? Well, Google's revenues were almost $110 billion in 2017, whereas Sinclair's were less than $3 billion. Sinclair has a lot of TV stations but doesn't come close to impacting our lives the way Google does. So Goodfriend's argument is more red herring than real, and he had better up his game if he wants to save his client.

> Relative size of Google and Sinclair
> **Sinclair revenue in 2017: $2.7 billion**[38]
> **Google revenue in 2017: $109.65 billion**[39]

We do join with Goodfriend in believing antitrust enforcement has become a joke in America, with both Republican and Democratic administrations ignoring the growth of a new kind of monopoly threat. We'll talk more about antitrust in Section 3, but for now we have to view Goodfriend's whistling as him begging for cover. Here's a former senior Clinton aide getting paid a bundle to circle the wagons for Google. He must feel embarrassed for himself. So embarrassed that he frames his retainer as a fee for doing battle to save America from the imminent alt-right takeover.

Creating a New Religion That's Above the Law

If there is a universal head nod to savoir faire, it is to the Apple logo. It makes tens of millions of people feel sexier, prettier, more virile, and even closer to God. Like a religion, Apple has its own belief system, its own idols, and a cult of people eager to worship. Among its congregants are the creatives, style-setters, and innovators. Instead of a rosary or the Book of Common Prayer in their hands, they have the iPhone—establishing a connection in the church of cool.

And with its line of products elevated to such exalted status, Apple is able to price at a premium—like an automotive company with Ferrari's margins and Honda's volume, making it the most profitable company in history. So profitable that just the cash the company has on hand very nearly tops the GDP of the entire country of Malaysia.[40]

And here's another way to look at Apple…

In December 2015, in San Bernardino, California, a young couple showed up at their office holiday party in ski masks and fired seventy-five rounds from AR-15s at their co-workers, killing fourteen before dying in a shootout with police. The FBI obtained the iPhone 5c of the male shooter, Syed Rizwan Farook, then obtained a court order for Apple to unlock the phone. Apple CEO Tim Cook defied the order.[41]

Cook has often communicated in speeches: "Our view of privacy started from our values, and then we created our business model to that; we've felt strongly about privacy when no one cared."[42]

Does this mean a company can choose to circumvent or ignore lawful investigations of crimes?

At the time, Apple argued that they could not permit the FBI to have a back door to their new operating system (iOS) because then it would be vulnerable to every Tom, Harry, and FBI Dick to come along. They also said that the government cannot conscript private companies to spy on private citizens.

The most entertaining retort to Apple's arguments came from NYU professor Scott Galloway in his excellent book, *The Four: The Hidden DNA of Amazon, Apple, Facebook and Google*. He wrote:

> "If Apple was creating a back door for others to use, it was a pretty unimpressive door. More like a doggy door. Apple estimated that it would take six to ten engineers a month to figure this out. That ain't the Manhattan Project. Apple also maintained this key could end up in the wrong hands and prove hugely dangerous. We aren't talking about the microchip that gave rise to the Terminator, which travels back in time to destroy all humanity. And the FBI even agreed to let the work take place on the Apple campus, ensuring it didn't become an app we can download from the FBI's website…
>
> "Their second argument, that a commercial firm shouldn't be enlisted in government fights against its will, is a marginally better one. However, does this mean if Ford Motor can construct a car trunk the FBI can't unlock, where it believes there is a kidnap victim suffocating, then the Bureau can't ask Ford to help them get in? Judges issue search warrants every day. They comply with search-and-seizure laws that prevent indiscriminate searches, and order homes, cars, and computers searched for evidence or information that might prevent or solve a crime. Yet, somehow, we've decided the iPhone is sacred. It isn't obliged to follow the same rules as the rest of the business world."[43]

Clearly in veneration of Apple's achievements, we have allowed a single company to rise above the law, no longer contained by it, unfairly disrespecting other companies which must play by the rules. Do you imagine for a second that if the shooter's phone had been an Android, Google would not have been forced to open it up? Or if it had been a BlackBerry, that congressmen wouldn't be calling for the Canadian government to insist the company break it open?

Instead, Apple is given a pass, creating two levels of companies in this country—those that must submit to lawful authorities, and those imbued with such religious gravitas that they are granted unequal stature. In short, they are granted the very inequality that their admirers profess to disdain.

Yes, Apple has given us an historic run of innovation and profitability, delivering one eye-popping billion-dollar product after another. But should that lift them to some exalted status above a nation of laws?

Putting Children in Harm's Way (Purposely)

A young student at Rutgers University named Tyler Clementi leaps to his death from a bridge after a private moment of him kissing another man is posted online in a hateful way. School officials and authorities decry the terrible posting, but no mention is made of the role Facebook, Twitter, and Apple's iChat play in this eighteen-year-old's death.[44]

Beautiful young Phoebe Prince hangs herself after nine teenagers stalk and harass her for months, calling her an "Irish slut" and "whore" on Facebook and Twitter as well as on the schoolyard. The online bullies face criminal charges; the social platforms are left in the footnotes.[45]

School bullying is nothing new, of course, but it's magnified online. The social platforms make it too easy, by design, and as a result, half of all teenagers say they've experienced digitally abusive behavior.

As far back as 2011, the American Academy of Pediatrics was warning that teens who use social network sites, and spend time comparing their profiles to their classmates', are more vulnerable to anxiety and what the doctors call "Facebook depression."[46]

This isn't just an American phenomenon.

A study in the *Archives of Pediatrics & Adolescent Medicine* of a thousand Chinese teenagers found remarkably similar results. "Heavy users" of social media were more than twice as likely to feel depressed as "normal users." Other factors are at play, the researchers noted. But the social platforms—and especially Facebook—have been negatively impacting the mental health of teens from the day the platforms were thrown by the chaos monkeys out into the marketplace (without a thought to the consequences for society).

Then there is the story of the young Indian man who posts a video on Facebook at seven p.m. saying he is about to commit suicide and asking viewers to share it. Moments later, he turns on the camera so viewers can watch him tying the rope around his neck and hanging himself from the ceiling fan. Both videos are viewed and shared thousands of times.[47]

His live suicide is the first of many to come, each one hosted by Facebook.

It's a dismaying fact that a small percentage of people will take their own lives. In the US, about one in ten thousand commit suicide. Most of them do so because, they say, they can no longer tolerate the stresses of life. Does that mean society should accept these stressors as something we can do nothing about? Or perhaps the stress agents should be treated like the accessories to murder they may well be?[48]

> **Nina Stanford, a media consultant, on how difficult it is for regulators to keep pace with the social platform's operations: "It's like we're ants in a hurricane."[49]**

A man who knows something about this is Dr. Howard Gardner, a professor at Harvard, who developed the idea of multiple intelligences—that is, the many ways to measure intelligence and the contributions that we can each make to society.[50] He calls the digital revolution we are going through an "epochal change," rivaling the invention of the printing press in its impact on how we interact with our fellow human beings.[51]

We've basically thrown our kids into this giant social petri dish from the moment we start posting adorable pictures of them in their pink and blue infant clothes:

- By age four or five, they are using a keyboard and playing games on colorful plastic digital devices.

- By seven or eight, the first set of kids get their own phone, diving into virtual worlds, staring at screens more than embracing the world around them.

- As early as six or seven, children may accidently see internet pornography.[52]

- In a few more years they'll be streaming videos and many will get Facebook pages—even though they're legally underage, and usually with parents unawares.

- By this time, at about age ten, they'll be living almost completely in a virtual world—putting on shows, taking off clothes, acting out and

performing for invisible audiences, with hardly a thought to the consequences of it all.

SurveyMonkey and *Fortune* polled 3,000 people aged eighteen to twen-
ty-four and found they "are the most apt to be online constantly" but also the
least likely "to think about their personal privacy" when using the web.[53]

This is a portrait of the youthful mind we all recognize. But it is also more.

This same polling also found that 67 percent of these young adults distrust
Facebook, 56 percent distrust Google, and 41 percent distrust Apple. So the
distrust is high, but so is the continued use. The simple definition of an addiction is not being able to stop.

And they clearly cannot.[54]

This puts the social platforms in the same category as Schedule II and III
drugs like opioids and cocaine—exceedingly addictive. We regulate the heck
out of Schedule II and III drugs, but not the social platforms. Yet each exerts a
related effect on the human brain.

People become addicted to the social platforms by design—very careful
design. Features are built in to induce certain behaviors: keep pushing that
button, keep scrolling on the page, life will get better for you. These are the
designs built in.

One Snapchat feature, for instance, assigned a fireball graphic next to the
names of all the user's friends. And it kept track, with big flashy numbers, of
how many days in a row the user sent a "snap" (a photo) to that friend; skip a
day and the number went back to zero. It was a subtle way to keep the users
hooked. And it worked like gangbusters.[55]

Even the mundane act of "posting" is addictive in nature.

Diana Tamir, an associate professor of psychology at Princeton, and Jason
Mitchell of Harvard's Neuroscience Lab studied this and published their results
in the paper "Disclosing Information About the Self Is Intrinsically Reward-
ing."[56] They ran MRI scans on people while they posted and found that even
posting something as simple as their dinner plans lit up the pleasure centers
in the brain. Lit them up just as brightly as they do with primary rewards like
food and sex.

It's no wonder that a poll of 1,000 teens by Common Sense Media found
them choosing to communicate with each other by text, social media, video
chatting, or phone 66 percent of the time, and in person 32 percent of the time.

For them, actual human interaction is becoming a relic of their parents' era. One commented:

> *"I start to wonder, are we getting into some negative feedback loop? You're distracted with people when you're with them, and they're distracted, and it isn't as fun in person so you'd rather be communicating online."*[57]

No study is required to prove how addictive these little behavior modification tricks have worked. James Steyer, a professor at Stanford, has seen firsthand the impacts of the social networks on his students in the lecture halls of a top institution:

> Technology *"has literally changed the way people relate to each other, get together, and present their image to the world."*
>
> *More than half of them say "they wished Facebook didn't exist."*

These very bright young adults didn't like how it engulfed their time, diminished the quality of their friendships, and damaged basic communications skills. But...

> *They had to be on it "because everyone else was."*[58]

Steyer talked about his own eighteen-year-old daughter, and how she just rolls her eyes when grown-ups natter on about the positive vibe of Facebook. It's very different for most kids. Facebook is being used not *in addition to* face-to-face relationships, but *instead of*. Posting and text messaging are quick and efficient, yes, but also cold.

Gone are the emotional nuances of facial expressions, the tone of voice.

Gone is the intimate act of looking people in the eye, or even hearing their voice over the phone, and knowing instinctively and through learned behavior what they truly are saying.

The effect can be cruel and damaging—especially for a generation that has never known any other way to communicate.

The supposed efficiency of texting and online messaging has stripped human emotion out of social communications.

The only real emotion comes from emojis—and though easily laughed at, they oddly help restore humanity to online communications. So much so that entire movies are being made with emojis as the stars. And we can't help but think this suggests a pining for some deeper connection that has been lost. But as a result, teens now spend an inordinate amount of time trying to read between the lines and wondering what the sender really meant, often undermining the very efficiency the texting supposedly was intended to communicate.

As a result of this, today's teens report feeling greater insecurity and social anxiety than their parents reported. When the most intimate communications—personal confessions, breakups, jealousy—are now communicated solely over a cold platform, the net result is a culture that's less compassionate and caring. This is not just something kids are going through on their own. Talk to kids honestly, and they'll admit that when they see their own parents "living" on their cellphones, they feel less loved and cared for.

Half Their Lives Spent on a Platform

Kids now spend more time on social media than anything else. More time than they spend with their families, their friends, school, exercising, daydreaming, eating and drinking, reading, chasing butterflies, even sleeping. In short, it's the biggest part of their life, with an average of nine hours a day given over to it; twelve hours if you count multitasking.[59] So, basically, half their lives. During this "half their lives" period…

What kinds of life lessons are they learning and who, or what, is most influencing their thinking?

What are the implications for healthy development when half your life is being played out on a public platform where there are huge social pressures to project an idealized image?

When every online action is designed to get a reaction, how do issues of self-esteem, narcissism, anxiety, and authenticity play out for today's teens?

For the first decade of social media, we simply couldn't answer these questions with any certainty. We had ideas, sure. But they were mostly anecdotal. Now, as time has passed, academics and researchers have been able to weigh in. Leading pediatricians, cognitive researchers, and child health experts have observed in clinical settings and in studies the impact that a constant digital connection has had on an entire generation of youngsters who appear to be having more problems with attention and concentration. Before, we didn't know for certain; now we do.

Researchers Can Now Quantify the Damage Done

MIT's Sherry Turkle was the first to sound the big alarm—long before anyone wanted to listen—about how the amazing technologies cascading out of Silicon Valley were increasingly isolating people in digital pods. Her seminal insights brought out in her book, *Alone Together: Why We Expect More from Technology and Less from Each Other,* said it all in the title. Our being constantly connected online has had the effect, rather paradoxically, of leaving us feeling more and more alone.

Dr. Dimitri Christakis, the director of the Center for Child Health, Behavior, and Development at Seattle Children's Research Institute, has found that attention deficit hyperactivity disorder (ADHD) has become ten times more common in children in the past two to three decades. ADHD is the term given to abnormal levels of distractedness, impulsiveness, and overactivity. And Dr. Christakis's studies link this sharp spike in ADHD to the design of social media:

> *"[Kids] can create a habit of mind where [their] brain is constantly seeking something more interesting, something more stimulating, because it's always available—and that leads to distractibility."*[60]

Texting and Facebooking may be more psychologically addictive than physically addictive, but addictive nonetheless. Christakis adds:

> *This addiction grows out of "a combination of genetic predisposition, coupled with exposure to those behaviors."*[61]

So some of us are more genetically predisposed to addiction and more likely to be hurt. And society has not figured out how to deal with this.

David E. Meyer, a psychology professor at the University of Michigan and chair of the Cognition and Cognitive Neuroscience area of the Psychology Department, has chronicled a steady decline in students' attention spans year over year as social media distractions pile up and linked the increase in these distractions to a steady decline in the brain's processing power.[62]

For two years, researchers tracked about 2,500 tenth-grade students and their usage of social media, games, and streaming video. In results published in July 2018 in the *Journal of the American Medical Association,* teens were found to have twice the risk of ADHD (such as difficulty completing tasks or remaining still) as a result of their online behavior: [63]

- Teens who used social media lightly had a 4.6 percent rate of ADHD symptoms.
- Teens who used social media heavily had a 10.5 percent rate of ADHD symptoms.

Researchers did note that correlation does not prove causation. Certainly other factors, such as lack of sleep, stress at home, and genetic history, played in. But this was the first longitudinal study to provide supporting proof that heavy users of the social platforms are twice as likely to have a life beset by ADHD.

**Corroborating findings from Dr. Jean Twenge's book,
iGen: Why Today's Super-Connected Kids Are Growing Up Less Rebellious, More Tolerant, Less Happy—and Completely Unprepared for Adulthood—and What That Means for the Rest of Us:[64]**

- **Eighth-graders' risk for depression jumps 27 percent when frequently using social media.**
- **Kids who use their phones three or more hours a day are much more likely to be suicidal.**
- **Teen suicide rate in the US now eclipses the homicide rate, perhaps because of smartphones.**

Another unrelated study by the US National Institutes of Health found that women who smoke or use drugs, or who are exposed to lead paint during pregnancy, have children with an increased risk of developing ADHD, as well. Ipso facto, frequent use of the social platforms has a similar impact on the child's body as smoking, doing drugs, and exposure to toxic lead during pregnancy.[65]

Our online behavior has now been shown to have direct neurological consequences. But was it any surprise, really?

Stare at a lake for twelve hours a day and you've got water on the mind; stare at a screen all day and you've got a billion bytes of data shaping you.

Nicholas Carr put this much better in his book, *The Shallows: What the Internet Is Doing to Our Brains.*[66] In it, he asked a simple question: "Is Google making us stupid?"

Carr's answer: Our sudden immersion in this new social realm has rearranged everything, from how we think to how we work and play. As he puts it: "Calm, focused, undistracted, the linear mind is being pushed aside by a new kind of mind that wants and needs to take in and dole out information in short, disjointed, often overlapping bursts—the faster, the better."[67]

In short, a pattern of addictive behavior.

With a mounting body of analytical proof and case studies showing the dangers of social media overuse, even the Big Tech Tyrants themselves, or at least the more honest ones, have begun to admit the faults of their systems and regret their years of work.

Bill Gates and Steve Jobs Raised Their Kids Tech-Free

In fact, the biggest names in tech—Bill Gates, Steve Jobs, and many more—raised or are raising their children tech-free, or darn near.[68] They are sending their children to Waldorf schools, which generally forbid electronics. They know what they've done. They know they've built behavior-modification empires, and they don't dare subject their own children to it.

In the words of Sean Parker, the first president of Facebook:

"We need to sort of give you a little dopamine hit every once in a while, because someone liked or commented on a photo or a post or whatever... It's a social-validation feedback loop...exactly the kind of thing that a hacker like myself would come up with, because you're exploiting a vulnerability in human psychology...The inventors, creators—it's me, it's Mark [Zuckerberg], it's Kevin Systrom on Instagram, it's all of these people—understood this consciously. And we did it anyway...It literally changes your relationship with society, with each other...It probably interferes with productivity in weird ways. God only knows what it's doing to our children's brains."[69]

He is echoed by Chamath Palihapitiya, former vice president of user growth at Facebook:

"The short-term, dopamine-driven feedback loops we've created are destroying how society works... No civil discourse, no cooperation; misinformation, mistruth... I feel tremendous guilt. I think we all knew in the back of our minds—even though we feigned this whole line of, like, there probably aren't any bad unintended consequences. I think in the back, deep, deep recesses...we kind of knew something bad could happen...So we are in a really bad state of affairs right now, in my

opinion. It is eroding the core foundation of how people behave by and between each other. And I don't have a good solution. My solution is I just don't use these tools any more. I haven't for years."[70]

This kind of honesty doesn't extend to the top of Facebook. Zuckerberg admits only to making some mistakes, otherwise saying basically, "The world has changed, deal with it."

Over at Google, the first real honesty came from their former design ethicist, Tristan Harris. He became so angry and disillusioned with Google, he walked away and then founded the Center for Humane Technology. There, he spilled the beans on some of the dark arts and algorithm-driven trickery used to keep people glued to Google's pages.

One example, particularly wicked and heinous because it preyed on young and underage teen girls, is the method Google used on YouTube when those girls went searching for diet videos. A girl would enter search terms in an effort to learn how to lose weight, but YouTube would instead recommend videos about anorexia. In our view, the recommended videos were slanted to give the user information about a disorder per YouTube's agenda, rather than what the person really wanted—which was just dieting information. Harris said these videos were presented as the top options because they better held the girls' attention, keeping them watching longer, only coincidentally (if at all) giving them what they asked for.[71]

A handful of people working at a handful of tech companies steer the thoughts of billions of people every day.

—TRISTAN HARRIS, former Google design ethicist

Mr. Harris says YouTube is "designed with a goal, to capture human attention," however best that can be accomplished. So while design engineers like him never really meant for their platform to spread conspiracy theories, for instance, those conspiracy theories were spread in mass—*because Google's recommendation algorithms determined that conspiracies keep eyeballs glued to screens.* And that's how Google makes its money.

Most of the tech leaders have joined this delayed-response chorus now, if only in an effort to save face. They are insisting that yes, they are mindful of the risks of too much screen time. And yes, they are launching very real measures to cut back on the usage of those who are liable to overload. But how they intend to accomplish that—if they even can—is still far from clear.

Meanwhile our nation's schools are on a drive, seeded by investments from the tech sector, to train kids to be more digitally adept and make even greater use of their devices. Nobody of good conscience, and particularly not us, would argue that our grade school curriculum should be devoid of digital. As in all things, balance is the key.

But that balance has been lost.

Teens Not Getting Enough Sleep to Function Properly

We know the importance of getting a good night's sleep—no matter the age. Good sleep is even more crucial to the physical and emotional development of children. Nobody argues against this. But the tech world has gone silent on the troublesome truth that digital devices are interfering with healthy sleep patterns.

Teens seldom stray from their cellphones. An overwhelming 71 percent say they usually sleep with or next to their mobile phone, with 3 percent of them sleeping with device in hand.[72]

Many text each other into the wee hours, when they should be sleeping, fidgeting nervously in expectation of their friends texting them back. As many as one third of the 100-200 texts that the average teen will send each day will be sent after the lights are out at night.[73]

Of course, the kids are also playing digital games before bed, also disrupting sleep as they are bathed in blue light. The shorter wavelengths in blue light are believed to cause the body to cut back on melatonin production and quash the delta brainwaves that bring on sleep.[74]

This is the kind of behavior responsible social platforms should try to bring into level balance, to the extent possible. But the platforms have no history of doing so.

Similarly, as far back as 2010, when the nation was becoming all too aware of the dangers of distracted driving, the social platforms did nothing to discourage texting while driving. It took a Pulitzer Prize-winning story from Matt Richtel of the *New York Times* to drive legislative efforts across the nation to ban the use of cellphones while driving and to require hands-free headsets.[75]

And yet today, nearly a decade later, with multiple states adopting safer driving legislation, with countless public awareness campaigns behind us, with a full 94 percent of drivers saying they support a ban on texting while driving, you can still drive down any street in America and you know you'll see someone texting while driving—alarmingly oblivious to the danger to themselves and other drivers and pedestrians.[76]

This obliviousness leads to more than 1.6 million crashes and 3,000 unnecessary deaths each year, according to the National Safety Council.[77] Unnecessary, because the device makers could easily make it difficult to text while a vehicle is in motion…but they refuse to do so and refuse any responsibility for the deaths, as well.

But should they make it more difficult to text and drive, and if they continue refusing to, should they be considered a party to these deaths?

Does Facebook Aid and Abet Teen Suicide?

In the spring of 2018, a team of Facebook engineers went on "lockdown"— their shorthand for "all hands on deck, we're taking fire." And they were. People were using their newly launched Facebook Live video-streaming service to kill themselves on the world stage.

This wasn't the first time the engineers had convened over this highly sensitive issue. Announcing a suicide date has become hideously common on the platform. Facebook doesn't reveal how many attempts at suicide their platform midwifes, but according to the World Health Organization, nearly 2,700 people kill themselves every day.[78] And with the Facebook user base now stretched to 2.2 billion globally, it's reasonable to conclude that dozens of people are announcing and filming their deaths on Facebook every day.

In this lockdown, as reported by Michal Lev-Ram in *Fortune*, Facebook was doing its best to both help out individuals who may be in trouble while also trying to reduce the chance of an outbreak of copycats. It's a fine line they walk at Facebook—because of their sheer size. Facebook has pledged to remove any live video or post that "encourages suicide or self-injury, including real-time depictions of suicide." But, just as importantly, they must be careful not to remove the content before "there is an opportunity for loved ones and authorities to provide help or resources."[79]

This kind of crisis control is obviously tough. It's the reason Facebook brought aboard 3,000 additional content moderators early in 2018, charging them with the task of sifting through the videos and posts of people who may be in trouble. Facebook has also developed "fast forward" tools that help moderators speed through more content, along with AI systems that can detect suicidal posts, and "heat maps" of audience spikes that might indicate something bad is about to happen.

This risk-control operation is simply unprecedented in history—owing to the sheer size and scale Facebook has grown to. As the world's gigantic marketplace, it's a big public stage with a billion people watching—or so it might seem to a deeply troubled person seeking attention. And that is why its size has become an untenable problem.

If the platform was less of a megaphone, it doubtless would encourage fewer of these terrible situations.

The Loss of Privacy: Why Our Children Are at Risk

A thirteen-year-old girl takes a nude photo of herself with her cellphone and sends it to her boyfriend. The two break up and the boy passes the photo around class. It's labeled "Ho Alert!" Within hours, the entire school has seen it. For months, the young girl is ridiculed and taunted, online and offline. She falls into a deep depression. She changes schools, twice. Her ex-boyfriend is handcuffed, jailed, and charged with distributing child porn—a felony count. Two obviously foolish and naive middle schoolers have their lives wrecked, with no way of reversing it.

This is a true story, and sadly, we know there are many more like it happening every day at schools across the country. We have new technologies being tossed into students' lives, and little (if any) of that quaint notion we once called adult supervision. Instead, we have thrown the bulk of the technology problem-solving burden on our children. Though they lack the maturity and experience to handle it on their own, we've effectively yoked them to the task, pitting them against algorithmic wizardry in the lopsided battle of the millennium.

For the full decade that Facebook's platform had been used or, rather, abused by kids doing these terrible things, Facebook cared only how their data fared.

Practically a child himself, Zuckerberg either didn't understand or didn't care about the welfare of young people on the receiving end of lots of good, and lots of bad. Facebook was basically saying, "Your problems are not our problems."

Google's approach was no different.

In an interview with CNBC, Google's CEO Eric Schmidt was asked about the impact of their search results on people's right to privacy. Schmidt thought about it and said, "If you have something that you don't want anyone to know, maybe you shouldn't be doing it in the first place."[80]

In fairness, Schmidt *did* walk that quote back. And he may be a fine man, not the callous ogre that quote would suggest. But the revelation in his words is the attitude and ethos that pervades Silicon Valley. They've ripped apart America's long-held commitment to individual privacy, not necessarily because they are evil, but because they are focused with laser intensity of making money... even if that means ignoring the truth and consequences of their actions on children and families.

Like mafia families or highwaymen of yore, they may be good people in private—perhaps even church-going and puppy-loving—but even their priests and puppies had best get out of the way when they plot the ruthless steps they'll take to build their empires.

They wouldn't dream of subjecting their own children to the dark side of the social platforms, but they can dissociate and feel no remorse at the impact on other people's children. This split personality is justified by such anodyne maxims as "data is virtue" but it's truly the single-minded focus on profit that trumps all other concerns. "Move fast and break things and people, and apologize later" has become more than a cheeky mantra. It has become a business model.

> *People on Facebook say "like me," people on Twitter say "listen to me," people on LinkedIn say "hire me," people on YouTube say "watch me," people on Pinterest say "show me," people on Instagram say "look at me," people on blogger [sites] say "agree with me," and people on Tumbler say "accept me."*
>
> —LEE HUMPHREYS, *The Qualified Self*[81]

> *Engineers running these platforms say "do what we want you to do."*
>
> —THE AUTHORS

Subverting America's First Freedoms

We've had the right to basic privacy for so long, it's easy to forget that it's one of the fundamental rights in American history and culture, often upheld over the years in decisions up to the highest court.

The maxim that "a man's home is his castle," an article of faith in English law, informs the heart of the Fourth Amendment to the US Constitution—that citizens have constitutional protections against arbitrary searches and seizures of their personal (and private) property.

Our legal system has long enshrined privacy as a fundamental right. And in the strictest legal sense, corporations have never been allowed to trample on these rights simply because they've invented some new product and wanted to sell it. Yet that is precisely what Silicon Valley has done over the last decade.

When youngsters get a cellphone today, whether it's Apple or Android, a whole series of little privacy invasions begins. First an account is set up on Facebook, then a music service like Pandora or iTunes is added, along with some game apps. Already they've handed over enough information so that advertisers know what to sell them and, worst case, so that predators know where to find them. As well, they probably handed over a parent's credit card information to buy a few of the apps.

All of this information can help create an amazing online experience, or the opposite. Either way, you can be certain that few youngsters stopped for even a second to think about their "privacy" or how their private information can now be used to exploit them. You can be equally certain they didn't read a word of Facebook's privacy policy or run through the phone's settings menu to enable any basic protections.

Meanwhile their phones are loading up with "cookies" and other tracking technologies that have one goal: to get to know them well enough to begin shifting their behavior toward a corporate interest. Some of those interests are perfectly acceptable ones; others are not. But with the average child's phone loading up with more than 4,000 cookies soon after initialization, there are surely too many in the "not" category.

Skirting the Children's Online Privacy Protection Act

Since 1998, the Children's Online Privacy Protection Act (COPPA) has made it illegal for businesses to collect personal information on children under the age of thirteen. Information that *cannot* be collected includes their name, phone number, email, street address, and Social Security number. This is officially why a kid can't get an account on Facebook and other platforms until their thirteenth birthday. But as far back as 2011, *Consumer Reports* had found that 7.5 million American kids under that age have been using the site.[82] It's literally child's play to create a Facebook page with a false birth date or identity.

And at such a tender young age, they are not only exposed to a barrage of adult themes, but are, by their youthfulness, much more vulnerable to problems of malware, virus infections, and identity theft, along with more serious threats such as privacy hostility, bullies, and predators.

Even if they do enjoy a pristine online experience—the way early vision-aries of the internet naively fantasized it could be—they are still barraged by advertisers who routinely target them with personal messages tailored to the data being collected on them.

Changing How Men and Women (and Children) Approach Sex

Of all the addictions Silicon Valley has dialed up, pornography comes in at the top of the list. Online porn is so prevalent now, it is completely skewing count-less years of human progress toward healthy sexual relationships. The internet has altered human sexuality in myriad ways. It has normalized what men and women traditionally considered beyond the pale. The National Institutes of Health tell us it has caused millions of men to experience porn-induced erectile dysfunction as a result of chronic masturbation.[83] The term "sexual attention deficit disorder" has left the urban slang dictionary and become an ad campaign underwriting prime time TV programming.

Ours is not some prissy grievance over the existence of porn—it has always been with us, and presumably always will be. But this is no longer just a guy's stash of Playboys. This is about the rewiring of the human brain in a systematic way. So much so that abnormally talkative men are willing to talk about their online porn addictions. And these men confess that "switching gears from porn's fireworks to the comparative calm of ordinary sex is like leaving halfway through an Imax 3-D movie to check out a flipbook."[84]

Many men have been so expertly deluded that they now genuinely wonder why women in the real world cannot, or will not, measure up to the porn stars they see online. They begin to realize that even if they are married or in a rela-tionship, they have another woman or another partner—and it is porn. They are said to be "dating porn." And their date is hosted by the platforms provided by the Big Tech Tyrants.

All by design.

Producers of porn know that a dopamine-oxytocin cocktail is released in the brain during orgasm, acting as a "biochemical love potion" that's deeply addictive.[85] And these online cocktails have been unscrupulously dialed up to be as addictive as possible—the better to profit. If real-world relationships are ruined along the way, the Big Tech Tyrants say, that's just the price of progress. But is it?

If a toxin is dumped into a river, the EPA goes after the perpetrators. If E. coli is found in lettuce at the market, the FDA tries to run down the source and punish the farmer, processor, or shipper—if intent to harm can be established. So when the Big Tech Tyrants allow far-from-harmless pornography to run over their platforms, should they be held responsible for the damage it's causing?

Ask anyone at the privacy desks of the platforms, and they'll say the privacy genie is out of the bottle, that ship has sailed, or some other equivalent of "just deal with it, dude." But just because the ship has sailed, it doesn't mean the sails don't need trimming regularly. "Your child is going to look at porn at some point. It's inevitable," says Elizabeth Schroeder, who runs a national sex-education organization based at Rutgers University.[86] Since our nation's privacy laws to protect younger people, and older, have not been updated since Zuckerberg was still in middle school and social media really didn't exist, the time is clearly overdue for an overhaul.

Far from the so-called enlightened "anything goes" and "it's all good" paradises that early internet pioneers imagined, there is a growing unease and sense that these social platforms have become fenced-off farms—participation farms, in some expert views—where the people (just like farm animals) have their data harvested for the farm's gain.

As people have grown aware of this, the overwhelming support these "farms" once enjoyed is eroding away like the topsoil following a Midwestern tornado. A Guardian poll in 2017 bore this out: Less than one-third of Americans see Facebook as good for the world. Only 26 percent believe Facebook gives a hoot about its users.[87]

The brightest among us saw the soil-stripping dust storms coming much earlier. None other than Tim Berners-Lee, father of the World Wide Web, foresaw the dangers of the social platforms all the way back in 2008. He cried out to his fellow technologists: Build "decentralized social networks...more immune to censorship, monopoly, regulation, and other exercise of central authority."

Only now are we seeing how prophetic Mr. Berners-Lee has always been.

In approaching these technologies, trying to make the best decisions for society going forward, we must keep in mind that technology is neither good nor bad—nor is it neutral, as the technology historian Melvin Kranzberg opined.[88]

We need to act on behalf of the good (many people love the connections that are fostered on the social platforms) and against the dangers (many people are being hurt by the platforms). Figuring out how to handle this beast of a problem is not easy. If it were, it would have been done years ago. But it is becoming the central challenge of our times, because the addictions of these social platforms strike at the beating heart of America.

What's missing in Silicon Valley, what's most needed, is simple accountability. There are different ways to make this happen. But first and foremost, the engineers and visionaries in the tech industry need to understand that they are not living on some enchanted silicon isle. They are becoming increasingly liable for the digital destruction they're pushing across the continents.

They need to summon all the ingenuity they've shown in creating wondrous new products, and put that same talent to work limiting the negative consequences of their products.

Their unwillingness thus far to do so suggests a higher authority will have to show them the right way forward for the public good.

STRIKE 2

Stasi Tactics

Stealing Privacy Through Surveillance Capitalism

Every Day a New Horror Story Surfaces

Jennifer bought a condom at the corner bodega, and the sixteen-year-old's file was updated for a thousand advertisers selling products desired by sexually active young girls...

Monika is a happily married woman who made a slightly political post on Facebook that so enraged an unhinged stranger that the stranger accused her of having an affair—posting it all over the social networks. It was untrue, but Monika spent $100,000 trying to clear her good name...[1]

Dave was taking some guys to the shooting range, so he loaded up on extra ammo for all of them at the Walmart. Unbeknownst to him, a red-flag note went to dozens of databases monitored in Washington and algorithmically tapped in Silicon Valley...

Eileen was fleeing an abusive household and was careful to turn off her Android phone's location services so she could not be tracked. But still there were hundreds of companies that had captured her precise location, and some of them sold that sensitive data to interested buyers...[2]

Leigh Ann was a teacher—she left her cellphone on her desk, unlocked. A student snuck up, scrolled through her apps, and found a spicy photo of her in the files. He stole the image and posted it on social media. The teacher lost her job for this...[3]

Unseen marketers were given access by Facebook to a "private group" for women with a high genetic risk of developing breast cancer. These women shared highly personal information during these sessions, which they believed were private. But in fact, all the sessions were made painfully public...[4]

The going rate on the dark web for a Facebook account is $5.20. The dark web, as you may know, is the anonymous part of the web only accessible with special software. And thousands of online accounts end up there in the hands of identify thieves. Facebook told authorities it would never happen again in 2007, then again in 2011, again in 2014, again in 2016, and again in 2018—obviously misrepresenting the truth each time...[5]

All of these are true stories. As if right out of Dave Eggers's 2013 dystopian novel, *The Circle*, about a hypothetical social platform that no longer sounds all that hypothetical:

> You and your ilk will live, willingly, joyfully, under constant surveillance, watching each other always, commenting on each other, voting and liking and disliking each other, smiling and frowning, and otherwise doing nothing much else...Secrets are lies. Sharing is caring. Privacy is theft.[6]

Few of us understand how Orwellian (or Eggersian) our world has become—how little privacy we have left. Or what could happen to us when the most private details of our lives are constantly being surveilled and exploited by the social platforms. Or when everything about us is known. And used. Sometimes for our convenience and pleasure, other times for exploitation and ruin. Either way, we are losing control of something we once considered our most precious personal asset.

Only once it is completely gone will we realize how much we've lost.

SOCIAL PLATFORMS—the business model from hell
STRIKE 1: Use dark pattern programs to addict people to the platforms
STRIKE 2: Take legal liberties with users they happen to dislike
STRIKE 3: Create an algorithmic identity theft machine

They Know More About You Than You Think

Consider *Wall Street Journal* writer Julia Angwin's off-the-digital-grid experiment. Angwin spent a year trying to live her life without leaving a single digital trace. In her book, *Dragnet Nation*, she chronicled the tremendous difficulty of her endeavor.[7]

For starters, she made a point of not using credit cards. It was so difficult, she then created a fake identity to get a single card for those times when it was essential for her to remain on, or have access to, the grid. She used burner phones, turning them off when not using them, and replacing them often. She set up an encrypted email service in the cloud. She only used subscription services, none of which were supported by ads. It was an exhaustive process that took weeks and weeks out of her year. Her conclusions:

1. **The average citizen stands little chance at protecting basic privacy.**

2. **We're paying a bigger price than we know for the privacy we've forfeited.**

Cameron Kerry of the Brookings Institution describes the situation we're in more amusingly:

> There is a classic episode of the show *I Love Lucy* in which Lucy goes to work wrapping candies on an assembly line. The line keeps speeding up with the candies coming closer together and, as they keep getting farther and farther behind, Lucy and her sidekick Ethel scramble harder and harder to keep up. "I think we're fighting a losing game," Lucy says. This is where we are with data privacy in America today. More and more data about each of us is being generated faster and faster from more and more devices, and we can't keep up. It's a losing game both for individuals and for our legal system. If we don't change the rules of the game soon, it will turn into a losing game for our economy and society.[8]

But it wasn't amusing for Jessica Rychly, a teenager from Minnesota with a broad smile and wavy hair, as reported in the *New York Times*:

> She likes reading and the rapper Post Malone. When she goes on Facebook or Twitter, she sometimes muses about being bored or trades jokes with friends. But on Twitter, there is a version of Jessica that none of her friends or family would recognize. While the two Jessicas share a name, photograph and whimsical bio—"I have issues"—the other Jessica promoted accounts hawking Canadian real estate investments,

cryptocurrency, and a radio station in Ghana. The fake Jessica followed or retweeted accounts using Arabic and Indonesian, languages the real Jessica does not speak. While she was a 17-year-old high school senior, her fake counterpart frequently promoted graphic pornography, retweeting accounts called Squirtamania and Porno Dan.[9]

All of this was being orchestrated down in Florida by a little fly-by-night outfit with a fake address and a shady past. But nobody cared, and people trying to drive up their "social cred" forked over millions of dollars to this outfit—in a giant social platform fraud.

- Twitter is reported to have 48 million fake accounts, or about 15 percent of its total accounts.
- Facebook recently disclosed that it had twice as many fake accounts as it had previously admitted—up to 60 million bogus accounts.
- An estimated 3.5 million stolen accounts, each sold many times over, are run across the social platforms, with Twitter being in the lead.[10]

It's identity theft perpetrated on a massive scale—and the social platforms implicitly participate.

Privacy Was Understood, Thank You Greta Garbo

For the first two hundred years of our republic, the basic meaning of "privacy" was well understood, if not easily described. It was one of the first freedoms, one of the fundamental rights of every American, even if not always well-defined—with any ten people offering ten reasonable-to-them definitions of what "privacy" means.

In fact, its strength was that it was both flexible and absolute, open to personal interpretation while being intuitively understood by all. "I want to be alone; I just want to be alone," as uttered by Greta Garbo in *Grand Hotel* and in *Ninotchka* as well, resonated in the hearts of millions of movie audiences. Garbo's pleas for privacy are still high on the list of the one hundred most-quoted movie lines (number seven and number thirty, respectively).

The government's respect for our personal privacy still makes it a crime to purloin someone's snail-mail letter from a post box, or to reveal sensitive medical information (HIPAA). Privacy as envisioned by the Founders not only related to their lives and times in the eighteenth century; the notional importance of privacy in protecting individuals continued as the country moved into the 19th, 20th, and 21st centuries. It retained its ability to stretch into new times, to help us maintain a reasonable control over how our personal information is being shared.

This right is as close to a sacred trust as anything in this country.

And then came the social networks. Unbeknownst to most of us, that sacred trust was shattered and our long-held privacy rights took a sudden dip into the ether and were anesthetized, taking our identities down with them. Each time we mindlessly ticked an I AGREE box, we gave away a few more of the protections we once relied on, and even took for granted. We might as well have slapped a sign on our backs saying STEAL ME, and done it in a choice of five ways:

1. **Social Harassment**

2. **Government Intrusion**

3. **Corporate Surveillance**

4. **Individual Commodification**

5. **Personal Objectification**

The thing about these new privacy intrusions delivered by the social platforms is that it's easy to ignore them until you've been seriously victimized. Then it's too late for coulda, woulda, shoulda. At that point, you've become just another identity-theft headline. Another page-six story in the newspaper.

Though it's of little comfort to anyone whose history has been stolen, sold, and resold, and flung across the digital universe, there are some modest-to-strong redresses to help protect our identities from nefarious absconders:

1. **Rely on consumer groups to monitor your rights** (*weakest*)

2. **Dial up the controls on your personal information** (*neutral*)

3. **Switch over to privacy-based social platforms** (*stronger*)

4. **Take monopoly power away from the Tech Tyrants** (*strongest*)

Cybersecurity pros flee Facebook:
- **Fifty-five percent are advising customers to rethink the data they share on Facebook.**
- **Seventy-five percent are limiting their own use of Facebook, or avoiding it entirely.**

> - **Only 26 percent believe people will be able to protect their online privacy in the future.**
> **This from industry professionals.**
>   2018 Black Hat USA security conference survey[11]

Social Platforms in a Race to Own Your Face

If you've bought a new phone or computer recently, you've experienced the latest in facial recognition technology. Apple has FaceID. Microsoft has Hello. And though still a little buggy, they can make it as "easy as smiling" to unlock and log on to a device. No more trying to remember your password each time. Sounds great, right?

Well, when scientists split the atom and unleashed amazing energy to light up entire cities, they also unleashed the power to turn those cities into dust. Technology is neither all good nor all bad, nor is it neutral. It is always capable of doing something. Just what that "something" is depends on the motives and methods of those who control the technology.

And so we see Facebook and other companies taking an amazing new technology and using it for "possibly good" and "possibly bad" ends.

- What do we know about the data they collect every time they "make a match" of your face?
- How much personal information have you handed over to the never-sleeping computers to monetize to the fullest extent possible?

Theresa Payton was chief information officer for the George W. Bush White House and has wielded the double-edged sword of the surveillance state. She's concerned that Facebook already knows your face better than your mother does, and better than all the government snoops do, as well. As she told *Digital Trends*, she's equally concerned about how Facebook will end up using this technology:

> "There are a lot of really cool things that could come out of this technology, but recent history tells us we need to play out worse-case scenarios…We need to understand that new technology will always be released a year or two before we really understand the ramifications of securing that data, as well as the legal aspects of protecting privacy."[12]

Fair to say, we have no real clue about all the ways our lives will be impacted by the technology now being released globally.

Facebook makes a point of saying that scanning your face will help "protect you from a stranger using your photo to impersonate you." Certainly true. But what else will likely be true, based on what we know about the social platform's past?

One of the principal motives for developing this technology is Facebook's desire to monetize what's known as "augmented commerce." They're helping advertisers turn simple Facebook ads into augmented-reality experiences. In these, the advertiser can literally leap from the screen and create computer-generated images that interact with real-world environments, blurring the line between what's real and what's virtual in the things you see and hear, and even feel and smell. Right down to the addition of "optical sensors" that are monitoring your breathing and heart rate.

For some people this could be a fantastic experience; for others it could be very disturbing—especially in its early, buggy stages.

One of the patents Facebook has filed for this technology is straight out of *The Minority Report*, a Philip K. Dick story set in a dystopian 2054 where giant computers know so much about us, they can foresee all crime and stop it "before it occurs," and everyone has a "trust level." It is this information about you which determines what you can or can't do, can or can't buy. Dick intended it as an authoritarian warning.

Facebook, never one to heed warnings or consider unintended consequences (unless they can be eventually monetized), is testing this technology now. By scanning and tracking its users, they can decide what augmented-reality experiences should be created to maximize the advertiser's investment based on our moods, emotions, outlook, the whole ball of wax—for very little about us will be unknown.

Google is working to get into our heads as well. They're developing systems for identifying unknown faces in images using data scraped from Twitter, Facebook, Gmail, competitive websites—everything, really. Essentially stalking us online so they can drive up our purchase volume in Google-affiliated stores.[13]

We are not only rats in Google's maze; every turn—right or wrong—we make becomes one more sales potential data point. Our lose-lose is Google's win-win.

Google can already identify almost anyone by name just by scanning their photo—to about 90 percent accuracy. Their "Reverse Image Search" technology is far from perfect, but it can deliver a very good "best guess" about who is appearing in the photo. It's a way for someone, anyone, to put a name to someone they don't know—just based on the photo.

To get this information, Google automatically scans all of our online communications, social networking, calendar entries, and apps to guess identities. Just like the old search process—only instead of entering a term to get a result, you enter a photo. Along with the photo could come additional information such as occupation, group memberships, and much more.

Facial recognition is being deployed faster than society and lawmakers can even hope to understand it. Just as Gordon Moore predicted the annual doubling of computer capacity back in 1965 (a rate which proved uncannily accurate for almost forty years), emerging technologies like facial recognition are showing nearly exponential growth in much shorter cycles. Such rapidly evolving technology has the potential to interfere in our lives even more than phone tracking has.

As we write, only two states, Texas and Illinois, limit private companies from tracking people by their "faceprints."[14] Without federal-level involvement, the social platforms will find it easy to circumvent whatever roadblocks state or local governments attempt. It's not an even fight.

We are hardly proponents of heavy-handed government regulation—far from it. But a lack of basic regulations and oversight is proving worse with each new "unknown" algorithm being tried out on citizens without their knowledge or truly informed consent.

High-quality, internet-connected cameras are now everywhere—from our front doors to automobiles, from schools to liquor stores. How often is face-recognition technology being used in them? Nobody knows for certain.

Then there are the issues of bias built into the face recognition technology.

Thus far, it performs worse on women and people with dark skin. These biases could lead to misidentifications of some people, opening a new can of privacy invasion worms.

For this reason, Microsoft president Brad Smith has called on government and enterprise to work together to figure out the ethical best practices for using facial-recognition technology.[15] Specifically, what permissions and restrictions should be put on business, law enforcement, and our national-security apparatus to ensure the technology is being used to the benefit of all?

Facebook Patents 24/7 Spying Technology

Facebook's mission is to track every aspect of our lives—not only what we've done, but what we will do. Their latest patent is for something called "Predicting Life Changes" and it's designed to predict the major events in our lives, up to and including the day we will each die.[16]

By knowing the most intimate details of our lives, Facebook can serve us ads that are more likely to influence our behavior. And that means they can charge advertisers more for those ads. For example, being able to predict when a woman will become pregnant lets Facebook serve ads for baby clothes and strollers—not only to the woman, but to all of her friends and family.

This capability goes well beyond those creepy ads that follow you around the internet—appearing to know not only what you have looked at online, but also what you've talked about offline. These can be deeply unsettling and disturbing, but they are actually based on an older technology known as "retargeting" that tracks your browsing and clicking history, and combines it with external databases to create a profile of who you are and what you're most likely interested in. Yes, very unsettling.

But Facebook is going way beyond retargeting, taking a giant step further down privacy's slippery slope with a "prediction" capability. It's using AI to figure out what's going to happen in your life before you do, and then to begin subtly manipulating your behavior to maximize the role the social platform plays in the events unfolding for you.

Facebook's supporters will insist that the social media giant would never misuse such information. Uh-huh. We've seen how often Facebook has been forced to apologize for doing precisely what they'd previously insisted they would never do

And they are developing this predictive capability not for kicks, we presume, but for profit at privacy's expense. The tool could probably predict when Mr. Zuckerberg will have to troop up to Capitol Hill yet again to apologize for misusing the tool!

Facebook is also patenting an app that can spy on you through your phone's microphone.[17] The mic listens to the TV or internet programs you may be watching, or it may overhear conversations between you and a friend, so it can "suggest relevant content" based on what you have heard or said. Or at least that's the marketing pitch.

Once again, predictably, Facebook insists that it would never spy on users. But obviously an always-listening Facebook is set up to repeat the company's past "break trust and apologize later" tactics that they've gotten away with for so long.

Facebook attorney Allen Lo was left to defend the patent when tech journalists raised more than an eyebrow. Lo maintained that the patent was filed simply to fend off competitors. When pressed to explain what he meant by that, Lo said, "Patents tend to focus on future-looking technology that is often speculative in nature and could be commercialized by other companies."[18]

Riiiight. We wouldn't want to be Lo, stuck trying to justify a company that first stole online identities in order to objectify women before evolving into a company that manipulates people online to make money. But that is his job. And every time Facebook messes up, the consumer comes out with less privacy and less protection, and more chances of being exploited online.

Day by day, this picture gets murkier.

Writing about Facebook's new life-cycle prediction tool and its spying tool, the wonderfully acerbic freelance tech journalist Joel Hruska captured what we all must feel. In his June 2018 article in *ExtremeTech*, he wrote:

> Facebook tracks the things you say. It tracks the things you don't say. It tracks you when you aren't on Facebook. It ships VPN applications that double as spyware. It's signed secret data-sharing sweetheart deals with various hardware manufacturers you never agreed to share your data with. It's profited from the wholesale abuse of its systems by companies like Cambridge Analytica. Mark Zuckerberg has been pulling from the same "sorry" playbook for literally the past 14 years…
>
> Mark Zuckerberg has never believed in privacy, not least because violating yours has made him one of the richest people on Earth…
>
> Is it possible that Allen Lo is telling the truth? Absolutely. But at this point in the company's history you'd frankly be an idiot to believe him. At no point since its inception has Facebook given the slightest indication that it meaningfully cares about your privacy. In fact, the one and only thing you can bet on with regard to Facebook is that it's always about 15 minutes away from its next privacy-related scandal. Does anyone actually believe that a patent like this won't directly lead to 'FACEBOOK ADMITS IT GATHERED EVERYTHING SAID OUT LOUD IN HOMES FOR SIX WEEKS' types of headlines at some point in the future? Because if you do, you literally haven't been paying attention to the past decade of

Facebook disasters. This is what happens when you combine 'move fast and break things' with 'actively corrode the very concept of privacy and declare it a social good.' It's a feature, not a bug, at least from Facebook's perspective.

It's long past time to stop pretending that Facebook has or will ever treat user privacy with anything deserving of the phrase 'respect.' It never has. It never will…[19]

Sometimes, you don't even know which app is listening to you.

You're talking to a friend, and your cellphone is sitting on a nearby table. Seconds later, on your laptop, you see an ad for the thing the two of you were discussing. Creepy, right? And the chance of it being a coincidence is just too unbelievable. So which app is listening to you and serving up ads based on your conversations?

Could it be Facebook? Apple? Google? Amazon?

Yes.

Ask Amazon CEO Jeff Bezos about this, and he'll smartly deflect the question by saying that there definitely is an inherent conflict between privacy and security. It's the "issue of our age" he says.[20] Then, without skipping a beat, Bezos insists that Amazon will fight the government's every attempt to obtain personal information from its Echo and other watching devices. Okay. But then if you look, you'll find that Amazon is marketing these very technologies to private businesses and law-enforcement agencies alike so they can…obtain personal information on us. Amazon is trying to have its cake and monetize it, too. So far, they're doing so. Quite successfully.

Obviously, there are ethical and privacy concerns for these new technologies. But are those concerns being addressed in any public way?

No.

Dark Pattern Tactics Reminiscent of the Stasi

We don't believe Mark Zuckerberg has any love for the tactics used by the East German Stasi during the communist era—neither believe it nor intend to imply it. But imitation is the sincerest flattery, as they say. And the "surveil and control" operation run by Zuckerberg's team at Facebook has no closer historical parallel that we're familiar with.

The East Germans ran the largest secret police operation in history. They kept files on one quarter of the total population. They could open anyone's mail and listen to anyone's phone calls, and they ran an extensive network of informants. By the 1980s, an estimated one in fifty people were working for the Stasi, and everyday citizens feared what this secretive organization was capable of.

In *Dragnet Nation*, Julia Angwin wrote of visiting Berlin and speaking with a German historical records administrator, Günter Bormann.[21] She showed him how Facebook profiles our lives now. He in turn showed her how the Stasi tried to do their own social network mapping; but even with all their informants, they had a hard time constructing profiles even close to the detail since perfected at Facebook. He told her:

"The Stasi would have loved this."

The Devil Is in the Settings

What the Stasi would have loved most about Facebook was the trickery and underhanded deceptions they get away with. On its face, TheFacebook, as it was originally known, looked like a fun way for college students to rate each other, a mildly more sophisticated "Hot or Not?" Not such a big deal in the greater scheme…though from all the abuse and cruelty it leveled at women from the get-go, we're frankly astonished that more scorn has not been heaped on Zuckerberg by the #MeToo movement. But we digress. For when you agree to use the Facebook platform, you have to click a little box agreeing to the site's "Terms & Conditions." That's where Facebook's mischief begins.

Facebook has never wanted its users to really know the possible downsides to clicking the AGREE box. So they began early on to make it difficult to dig into it. In a practice known in technology design as "dark patterns," Facebook began designing their site to stealthily steer users toward Facebook's favored options, which many users would not have chosen if fully informed of the consequences.

Privacy Advocates Assail Facebook's and Google's "Dark Art" Practices

Facebook's and Google's privacy settings have grown more and more deceptive and manipulative over the years. By 2018, eight different consumer watchdog groups had petitioned the FTC to investigate the platforms for misleading and manipulative tactics.[22]

> **Every major privacy group is assailing Facebook's and Google's manipulative practices:**
> - Electronic Privacy Information Center
> - Campaign for a Commercial-Free Childhood
> - Center for Digital Democracy
> - Consumer Action
> - Consumer Federation of America
> - Consumer Watchdog
> - Public Citizen
> - U.S. PIRG
> - Common Sense Media

These consumer advocates represent both liberal and conservative constituencies—theirs was not an attack by one political camp or the other to score points. It was, rather, a culmination of years of worrying, reaching a bubbling-over point when the Norway Council of Consumers released a report titled, "Deceived by Design."[23]

This report came out of the most thorough investigation yet on how Facebook and Google manipulate users. It focused on how the social platforms trick users into accepting privacy settings that reveal personal information about themselves far in excess of what's needed to enjoy their services. Key excerpts from the report:

- The platforms employ "design, symbols and working that nudge users away from the privacy-friendly choices."
- The platforms are "circumventing the notion of giving consumers control of their personal data."
- The companies threaten users "with loss of functionality or deletion of the user account if the user does not choose the privacy intrusive option."

Facebook stops just shy of outright scare tactics in their onboarding process. They first make it sinfully easy to opt in to their facial recognition tool, for example, but annoyingly hard to say no. Users are given a bright blue box front and center that basically says, "Yes, scan my face for your files!" But if users want to decline, they have to click though to the "Manage Data Settings" page and go hunting for an option to decline the scanning.

What makes the dark-pattern designs cross the line from clever to cruel, insipid to insidious, is the ways they can shift your behavior in a thousand little nearly imperceptible ways. Small and subtle, one after another, guiding you down a road you might not otherwise choose.

For instance, when you come onboard Facebook, and Google as well, you find that all of the privacy options have been disabled by default. If you are not paying close attention, you won't even know you had a choice in the first place. You are constantly opting out of things, not opting in. If you want to enable your privacy options on Facebook, you must run through thirteen screens, versus only four if you are comfortable forfeiting all your privacy.

A similar thing happens if you go over to a Google page. There you'll get a rose-tinted explanation of how ad targeting works. They say it's designed to make ads more relevant to users. There is no real explanation of the benefits of turning off ad targeting. All you get, instead, is the warning that "You'll still see ads, but they'll be less useful to you."

If you want to disable Google's ad targeting, you will be further warned that you that can no longer mute some ads going forward. This can't help but confuse. Does it mean that if an ad pops up during a show you're watching, you can't mute it? Rather than continue with the mental gymnastics, if you are like most people, you just click the box—share my data! Google has darkly designed your acquiescence.

Again, these are Google's and Facebook's platforms. Theirs to monetize. But the larger they get and the more deeply rooted into our lives they become, the less control we have over our personal data and the more vulnerable we are to some bad outcomes. The social platforms are constantly insisting that we are the ones who are in control, but in truth:

It's becoming their world and we're just living in it.

Because of their mammoth and sudden reach into our lives, the social platforms have put us into something of a binary conundrum: privacy or sharing? During a stretch of six to twelve hours a day, depending on our media use, we are presented with a carefully curated set of nudges—a steady drip of convincers, shifting us into the "sharing" column (along with most everyone else).

What Is Being Done About These Privacy Thefts?

Facial recognition technology is just the latest flavor of socially sanctioned thievery, for that's what it is. Social platforms have been getting away with identity theft for a decade. Now, we'll admit that not everyone is offended by the thievery. Some truly like it. And judging from the attendances at TED Talks given by leading minds in Silicon Valley, they have a lot more credibility than leaders in most industries, certainly more than politicians or the media.

They have such a reserve of credibility, in fact, that they could have expected to continue getting away with some questionable business practices for a long time. But in 2018 we found out that the people do have a limit. The social platforms were put on notice in Europe and in America that it was time to fix their problems or face the regulatory strictures the medical community grinds under.

When we visit a doctor now, we must fill out HIPAA forms that detail the legal protections we enjoy. Some people—some patients, even—think these protections are over-the-top onerous. Others feel protected from a medical community or related contractors that often abuse their privileged access to the most intimate details of our lives. This latter group has prevailed, and, as a result, HIPAA regulations are enforced by the federal government.

Similar rules and protections are in place in the financial industry—again, quite simply, because of all the well-known, often-devastating abuses that have affected rich and poor alike. Now, with the social platforms abusing our personal data, how is their situation any different from that of medicine and finance, both requiring extensive regulation?

Surely a platform that addicts millions to it, gets a free pass on the free-speech rules all others must follow, and runs the world's first algorithmic identity theft operation, is a candidate for overhaul that's in the national interest?

These social platforms have been elevated to a central role in our lives, a trusted role. And yet they are an entirely new creation to us. Nothing like them has ever existed. There's never been an opportunity to modify behavior on the intimate level they're capable of. This is something that, as a society, we must understand better.

Take the latest iPhone, for example. If you own one, you probably know this about it: You can set it down somewhere nearby and simply say, "Hey Siri," and your iPhone will awaken and say, "Go ahead." You can then give it any number of instructions, which it carries out quite ably. But you also know that it is always listening.

And if there beats in you a human heart, you can think back over the past month when you've given voice to ideas you meant only for yourself—the "I wish I had…" comments, the business plans, the personal rants, the most tender expressions—none of which you ever intended for unseen engineers to capture and monetize.

It's the same with Amazon Alexa and Google Home, now sitting on the countertops in millions of homes—just listening to every word we say. Yes, they do listen for a keyword to turn on and take instructions, but that means they are always listening. Apple, Amazon, and Google all insist that their devices are listening only to you, and that no ongoing recordings are being made of what is heard.

But we've seen that we cannot judge the Tech Tyrants by what they say; rather we must judge them by what they have done. And looking at their short but spirited history of subtly deceiving users about the outcomes of clicking AGREE, we know our privacies are going to be compromised, we just don't know how harmful it will be.

> **The Problem Is the Social Platform Business Model**
> - **Problem 1: Social platforms built their business model on using personal information to make their fortunes.**
> - **Problem 2: Social platforms are not open and honest with their users about how they use this personal information.**
> - **Problem 3: Social platforms offer no alternatives—no real solution for the user who wants to protect their privacy.**

States Taking Action

Californians, who are closest to the Silicon Valley machinery, launched a ballot initiative for November 2018—to let state voters decide how best to protect against privacy invasions. The measure meant to:

- Let consumers "opt out" of the collection and sale of their personal data.
- Expand the definition of "personal data" to include geolocation, biometrics, browsing history.
- Allow consumers to pursue legal action for violations of the law.[24]

The very idea of granting Californians sweeping new privacy rights spooked Silicon Valley's tech denizens, and they went on the offensive. They convinced sponsors of the initiative to withdraw the ballot measure if the state legislature could pass some compromise privacy legislation, known as AB 375. So, behind the scenes the battling began, and when AB 375 finally passed both the State Assembly and the Senate without any opposition, Silicon Valley got what they wanted: toothless legislation they could continue ignoring.

The lobbying coalition that was convened to shut down the ballot measure included a who's who of the tech world: Ann Blackwood of Facebook, Lisa Kohn of Amazon, Mufaddal Ezzy of Google, Ryan Harkins of Microsoft, Walter Hughes of Comcast, and Kate Ijams of AT&T, along with reps from Uber, Verizon, Cox Communications, the Alliance of Automobile Manufacturers, and the Interactive Advertising Bureau. They had pledged to spend north of $100 million, if necessary, to keep everyone's hands off their business model.[25]

In the end, they only needed to spend a small fraction of that.

Not Just the Social Platforms, the Telecoms Too.

We have focused on the social platforms here, though the same concerns apply to the telecom titans—AT&T, Verizon, and Comcast, primarily. In fact, these companies have more of our personal information at their fingertips than the social platforms do, and they have it available on all of the computer screens in thousands of offices.

Telecom companies are the physical gatekeepers of the internet. They can monitor our behavior on our cellphones, cable and satellite TV, and internet browsing. Telecom companies can sell this user data to third parties, while Facebook does not. However Facebook does sell information to third parties that is derived from personal data. This way, Facebook can harvest more internally valuable information from users' News Feeds.

Nonetheless, the importance of securing privacy protections across every industry that holds our valuable personal data cannot be overemphasized.

Fake News, or Fake Medium?

Before Donald Trump even declared his candidacy for the presidency, he had begun pointing out how much "fake" had entered into our world. In that, he was shining a light on what had become the dirty little secret of the internet. And so we invite you to take a very quick scientific survey:

A. Sixteen percent of the internet is fake

B. Thirty-six percent of the internet is fake

C. One hundred percent of the internet is fake

If you had a hard time choosing between A and B, you're like most. Correct answer: B .[26]

So much of what you see on the internet is fake that if you spend any time there, you've crossed over to nonhuman time:

1. There are the explicitly fake nonhumans Alexa, Cortana, and Siri, and the billions of conversations these crafty bots are having at any given moment...

2. There's the Belstaff badass leather jacket you bought because it had a lot of good reviews, most of them generated by clever little elf bots...

3. There's the comedienne you found at the top of the search results—only there because of a Twitterstorm of fake followers she acquired by deploying armies of bots...

4. There's the online chat you're having with a Verizon customer service rep that's prompt and clear and totally automated...

5. And on and on it goes as high as you want to count.

These nonhumans have become an extension of our lives. If we are younger, they can even get attached to or rather locked into our peer identity. They then become an influence, the psychologists say, that will remain with us through our lives. So today's young people will be the first generation in history to have the influence of "fakefulness" loom larger in their lives than "factfulness."

As more things become fake on the internet—driven by AI-driven bots—we will, before long, leave behind the quaint ideas of "fake news" as we see nonhuman activity escalated to such levels; we now have almost entirely "fake mediums."

Bots Run Amok

The social platforms make loud noises about fighting back against all the fake accounts that infest their platforms. But it's all for show; they require those fake accounts to succeed, and they benefit from them—grandly.

Any decent employee of Twitter might, on some level, wish at night that their platform was bot-free. But it's the bots that amplify the activity and intensity of the service. Massively social bots can influence real people much more skillfully that most anyone imagined.

They can create separate social realities, like *The Truman Show*, for each one of us, and they can make a lot of money doing it.

They can also cause tremendous harm. Earlier we talked about Jessica Rychly, and how her social profile was stolen and sold across the social platforms—as if she was some kind of 21st century human chattel. Rychly's online profile is just one of an estimated 3.5 million accounts that have been stolen and exploited on the social platforms.[27] This thievery continues and indeed is flourishing today, largely because our media-obsessed world places such a high price tag on fake profiles.

Trafficking in Fake Currency

In fact, there are hundreds of thousands of people who buy these fake profiles in order to make it look like more people follow them online, and therefore, more people consider them opinion leaders, and finally, more people want to "be" them.

It's an entirely fake credentialing subculture dreamed up originally by some low characters, but now out in the open and flourishing. And for those many thousands now buying these fake accounts, they're not doing it just to appear popular. They're doing it because they are judged by how many "likes" or "followers" they have. They'll get paid more (actors' contracts are often tied explicitly to their Twitter follower count), or even land a better job (business consultants with tons of endorsements from "big names" on LinkedIn get hired faster at higher retainer fees). Now, many thousands of people's careers are entirely interwoven with these virtual status symbols, this fake narrative—so much so, they may even forget it's fraud, or, sadly, accept the fraud as the price of thin fame.

At least forget or accept until they are questioned for real by real reporters doing real stories. When in 2018 the *New York Times* ran an exposé on the theft and subsequent peddling of online accounts, correctly calling this a "black market," suddenly a lot of well-known personalities were seen putting on a show of outrage. "How could I have (unwittingly, unintentionally, untruly, etc.) bought all these fake followers—I had no idea!" Then they moved quickly to distance themselves from the Florida boiler-room operation that had been exposed in the *Times* article for spreading this counterfeit currency far and wide. "Distancing themselves" probably just meant going and buying the fake followers from another scammer outfit ready to step in and "fill a need," as they say.

Meanwhile, what becomes of people like Rychly who've had their identity stolen and resold over and over online? What redress does she have when the scam artists steal her profile, the celebrities buy it, and the social platforms host it?

None, now.

Google Is Pronounced Obama in Klingon

Are too many people locked into abusive co-dependent relationships with Google? Has Google been quietly aligning with the federal government to exert even more power? Should Google be left alone in the marketplace, or regulated as a public utility, or broken up?

Google's original slogan of "Don't Be Evil" will surely find its place among the most idealistic naivetés in the pantheons of corporate history, or amongst the greatest cons—it's too early to tell. We don't pretend to know Google's co-founders Larry Page and Sergey Brin well enough to decide between the two options ourselves. Or whether to choose both—them having naively run a con, if you will.

Or maybe the two brilliant Stanford students started out with the best of intentions, but soon found that running a technology company isn't all cardinals singing. Such as when, one day in 2010, they decided to throw their own policy book out the window and arbitrarily enrolled their users in a new service, Google Buzz, despite those users having opted out. Google opened up these accounts for users who had explicitly told Google no thanks.[28] As part of the Buzz experience, Google was also releasing the names of users' email recipients to the public. If you sent a private email to a friend, Google just made it public. It was astounding that any company would be so…lame.

For this massive privacy transgression, Google ended up in court. They got a quick and painless settlement and promised (a) not to misrepresent their data handling policies ever again, and (b) to be audited by authorities for a long twenty years of audits to ensure compliance. Thus far (as of 2019) into the twenty-year probationary period, Google has roundly ignored the court's order again and again.

Just two years after agreeing to play nice, Google went right ahead and evaded the Safari browser's cookie-blocking feature with its +1 button, allowing its engineers to track a user's ad preferences.[29] The FTC fined Google $22.5 million that go-round. And, oh, how it must have stung, representing, as it did, 0.04 percent of Google's $50.2 billion annual revenue at the time.[30]

By 2018, the company had been sued so many times by different regulators on multiple continents that they officially dropped "Don't Be Evil" from the company code of conduct, presumably because even they could no longer keep a straight face saying it. We heard that they replaced the slogan with "Evil Be Good" but were sued by the Chuck Berry Family Trust and withdrew it, though we could be wrong about this.

Public Concerns Reach a Tipping Point

The internet has so accelerated our perspective of time, it's easy to forget that only a few years ago we were marveling at the wizardry behind Google's search engine technology. Their original mission statement, "to organize the world's information, making it universally accessible and useful," was a heroic act of unprecedented scale. Finding almost any answer suddenly became as easy as breathing in and breathing out. Just ask a question and hit ENTER! It was so totally amazing, we were oblivious to the crossed lines, or we chose to forgive Google for a lot of line crossings.

But concerns were mounting.

And in 2018 we reached a tipping point—when everything began tumbling down, hitting the ground hard. There's blood, call an ambulance!

In January 2018, a *Wall Street Journal* feature story made a forceful case against Google for squelching competition, restraining trade, and usurping privacy or a level requiring antitrust action.[31] And the next month, the *New York Times* did the same."[32]

For the nation's two leading newspapers to reach the same conclusion, despite their moorings in ideologically opposed camps, meant the country was coalescing around the need for action at the highest levels.

Anyone who had been hoping to get action during the Obama years would have been sorely disappointed. If there was an alternative spelling or flip-side mindset for Google, it was Obama. From the day he first toured their offices in Mountain View and declared "we are one" to the day he left office, Obama took pride in embedding Google into the decision-making infrastructure of Washington.

There was the revolving door of crony capitalism, more accurately described as a spinning door, with hundreds of meetings (as well as hundreds of people) shuttling from jobs at the Googleplex to the Obama administration. More specifically, as *The Intercept* found in an investigation they headlined "The Android Administration," Google and government were peas in a pod "with 252 job moves between Google and government from Obama's campaign years to early 2016, and the 427 meetings between White House and Google employees from 2009 to 2015—more than once a week on average."[33]

Or, as Steven Levy said pithily in his 2011 book, *In the Plex*: "Google and Obama vibrated at the same frequency."[34]

There was also the lobbying prowess, with Google passing General Electric in 2017 to become the largest corporate spender on lobbying in Washington—$18 million a year, which buys a lot of allegiance.

Underpinning their shared policy agendas was a shared ethos, a mutual admiration and belief that society's challenges were best met with their own enlightened constructs about how folks should live their lives. Adam J. White, a research fellow at the Hoover Institution, described this Google-government symbiosis in a *New Atlantis* article titled "Google.gov" (and worth reading in its entirety):

> "Both view information as being at once ruthlessly value-free and yet, when properly grasped, a powerful force for ideological and social reform. And so both aspire to reshape Americans' informational context, ensuring that we make choices based only upon what they consider the right kinds of facts—while denying that there would be any values or politics embedded in the effort."[35]

Of course, this special worldview shared by Google and Obama's political appointments was dealt a serious setback with the change of administrations in 2016. Nonetheless, their "progressive" ethos remains intact at most levels of government. It will not be easily dislodged or course-corrected by members of the Trump Administration who wish to do so.

As if speaking to this very issue, Obama, in his post-presidency, has said Google should not be regulated or broken up. Instead, he believes "they are a public good as well as a commercial enterprise...shaping our culture in powerful ways."[36]

Obama's choice of words and his juxtaposing "public good" with "shaping our culture" struck many. What did he mean? The interpretation that White reached in his investigation was that Obama's time in office was over for him, but Google must go on advancing their shared progressive vision of the public good, with the search engine staying the new course of giving searchers the results they ought to have. And then White concluded that Google "will become an indispensable adjunct to progressive government. The future might not be U.S. v. Google but Google.gov."[37]

Some might think Google.gov—the literal taking over of government by Google—is far-fetched and silly. We might agree, but only to a point—for we have heard Bay Area partisans bandy about a new motto on the dollar bill: IN GOOGLE WE TRUST.

We do know that Google's inroads into the federal government, and thus ultimately into every bit of data defining our lives, would have increased exponentially if Obama had another eight years in power. Two peas in a pod, Obama and Google were committed to directing how people think, and the already vast reach of Google insured a much deeper penetration into our lives, privacy having long since gone out the window.

An Innovation-Killing Machine Masquerading as a Search Engine

Google's Vast Reach

- **Originally called BackRub (far from mind, apparently, was their future world domination)**
- **Conducted 1.3 trillion searches in 2018 with $61 billion in ad revenue**[38]
- **Controls 40 percent of all online ad revenue**[39]
- **Reorganized as "Alphabet" with products for:**
  - Internet search
  - Web email
  - News feed
  - Video hosting
  - Web browser
  - Maps
  - People tracking
  - Calendar
  - Social network
  - Office productivity suite

- Blog software
- Video rental
- Software store
- Mobile payment
- Language translation
- Video conferencing
- Cloud storage
- Cloud computing
- Mobile operating system
- Digitized books
- Internet TV
- High-speed internet
- Computers
- Home automation
- Robot assistants
- Self-driving cars
- Life extension
- Contact lenses
- Internet blimps
- Drones
- Wind turbines
- Skunkworks

Google's first big aha moment came from the perfecting of objective ranking metrics. It was a giant breakthrough—a big win that left 1990s-style search engines in the dust. But this "objective" approach also had an Achilles' heel. It could be gamed by web developers eager to improve their search rankings. So Google was forced, even if it didn't want to, to wrap its ranking algorithms in "black box" secrecy. This would never have been a problem to anyone if Google could be trusted to not manipulate the search results.

Turns out, they couldn't be trusted. But we didn't know it for a long time.

The first worries came from research conducted in 2010 by a Harvard business professor, Benjamin Edelman. In his studies of Google, he found what he could only assume was "hard-coding" of their search algorithms.[40] That is, the engineers were being told to override the automated algorithms and tweak the search results so that queries for certain keywords would bring up Google's own websites before those of others—no matter which was more objectively popular.

In stock trading, this practice is known as front-running. And in stock trading, it is not only unethical, it is illegal, because it is using insider information to cheat the unsuspecting. The good professor had proven that this was Google's new business model.

When the staff of the Federal Trade Commission saw the professor's findings, they launched their own investigation. And in 2012, they released their own findings on how Google regularly prioritized the websites it owned over those of its competitors, all while selling itself to the public as a company simply trying "to organize the world's information, making it universally accessible and useful."

When the FTC confronted Google with these charges, nobody at Google even bothered to deny them. In fact, Google senior exec Marissa Mayer had previously acknowledged at an industry gathering that it was company policy to arrange Google properties at the top of the search results. In her words: "It seems only fair, right? We do all the work for the search page and all these other things, so we do put it first."[41]

It would be fair, if Google were a library and the search results a Dewey Decimal classification of old. Google built it, Google benefits from it…the logic would go. Except for two things—two very important things, in fact:

1. **Google presented itself to the world as a neutral provider of information.**

2. **Google charged other companies to be placed high in these very results.**

Thing one and thing two matter.

1. **If Google is not a neutral arbiter of information, then how do they make decisions about the information they present to the world?**

2. **And if we've entered a time where one company's technology lets them front-run their competitors and there are no laws forbidding it, then the laws themselves have not kept pace with the technology.**

As it turns out, we'd soon have an answer to the first question—an answer that would persuade us of the need to resolve the second question.

And once again, it was Mr. Obama who would bring the matter into clearer focus.

In a 2007 speech to Google employees, after which CEO Eric Schmidt would endorse Obama and campaign for him, *Obama talked about the need to use technologies to counter all the wrongheaded thinking in America and to confront folks with real facts.*[42] His language was actually quite a bit more imperious, but he is an ex-president, and we see little gain in editorializing. We care only that the Googleplex auditorium erupted in cheers that day, for the assembled engineers knew exactly what Obama meant.

They knew he meant that folks at Google, like he himself, were the ones who are able to get their facts straight, the ones with a head on their shoulders, the ones enlightened enough to keep the people properly informed. The ones, the ones, the only ones. Evidencing a level of arrogance, it might be easy to laugh off—except that Google had by then moved into the position of being able to control what gets seen and said by a majority of people.

Only after leaving the White House did now-citizen Obama begin to chill on Google and wonder if their technology had, in fact, outpaced society's regulatory framework. In a speech to MIT's Sloan Sports Analytics Conference meant to be off the record, Obama talked about Google much more candidly and in fraught terms: Google is exacerbating "the balkanization of our public conversation."[43]

In other words, Google had moved, in Obama's mind, from being part of the solution to being part of the problem. In that, we can agree.

Google, like Facebook, was certainly listening to Obama's MIT speech. They are not deaf to the growing concerns of many, including the calls for its head. So Google has launched its own campaign to fight "fake news," among other things. Google has also begun tweaking its search algorithms to prioritize content considered "more credible." In Google's words: We'll deal with "the spread of blatantly misleading, low-quality, offensive or downright false information" by down-ranking "offensive or clearly misleading content" and up-ranking "more authoritative content."

Google has also added a "Fact Check" feature that tells viewers if the information they are looking at is authoritative and authentic. That is, authoritative and authentic in the view of the folks at Google and its fact checkers working for about a buck an hour in cubicles in Manila and Bangalore, and flying through stories as if in a flipbook to maintain their quotas.

Google's new search results have included fact-check boxes for results thought to include "claims." The inset box has displayed the claim that someone made, who was making the claim, who fact-checked it, and whether that fact checker found it to be true or false.

Away have gone the old "objective" algorithms that supposedly returned perfect results—or at least they are gone for topics that Google has determined are controversial, which could be a broad net.

Finally, Google has taken a step further with the Trust Project. Here they are working with loads of news organizations to come up with trust indicators that can be used to decide on which search results should be prioritized. These eight indicators are:[44]

1. **Best Practices: What are your standards? Who funds the news outlet? What is the outlet's mission? Plus commitments to ethics, diverse voices, accuracy, making corrections, and other standards.**

2. **Author/Reporter Expertise: Who made this? Details about the journalist, including their expertise and other stories they have worked on.**

3. **Type of Work: What is this? Labels to distinguish opinion, analysis, and advertiser (or sponsored) content from news reports.**

4. **Citations and References: For investigative or in-depth stories, access to the sources behind the facts and assertions.**

5. **Methods: Also, for in-depth stories, information about why reporters chose to pursue a story and how they went about the process.**

6. **Locally Sourced: Lets you know when the story has local origin or expertise. Was the reporting done on the scene, with deep knowledge about the local situation or community?**

7. **Diverse Voices: A newsroom's efforts and commitment to bringing in diverse perspectives. Readers notice when certain voices, ethnicities, or political persuasions are missing.**

8. **Actionable Feedback: A newsroom's efforts to engage the public's help in setting coverage priorities, contributing to the reporting process, ensuring accuracy, and other areas. Readers want to participate and provide feedback that might alter or expand a story.**

It looks comprehensive, right? So comprehensive that it might be something they'd teach in journalism school or you'd see in frames on a wall at a famous newspaper. Which suggests that Google has put themselves in a new business.

The news business.

In a mad dash to appear credible, Google has demoted the algorithms that originally made them great (or at least had done so, until Google's own management adulterated them) and is now promoting an unseen army of people working like slave laborers to parse fact from fiction based on Google's enlightened directives.

If this doesn't sound like a hopeless task, given that one man's fake news is another man's gospel truth…

If that doesn't sound like the potential weaponization of information…

For example, how does Google get their bearings right on issues on which intelligent people hold widely different views? Issues such as climate change and abortion, to name just two? The left glibly says they have all the facts on these, and everyone else are dummies. Like when Al Gore admitted to telling untruths—such as cities like New York would be underwater by the year 2010—to make his global warming platform for the presidency so persuasive to the public. Oh, wait, that's not a good example, is it?

Facts are not facts because someone says so.

Grass is green; fire will burn you; Trump is a flaming racist. A liberal believes all to be facts, when in truth two are facts and one an opinion, no matter how much evidence or reason is found to support it, because just as much evidence and reason can be found to negate it.

These liberals seem to have forgotten the Spinoza and Rooney they should have studied in college. "No matter how thin you slice it, there will always be two sides," wrote Baruch Spinoza. And, "People will generally accept facts as truth only if the facts agree with what they already believe," quipped Andy Rooney, also demonstrating that keen insights come as often from comics as philosophers.

So, how are the left-leaning engineers and quota-driven fact checkers at Google going to ferret out fact from fiction in the millions of new search results they process every minute? It sounds like an impossible task to toe a neutral line…if that was even of concern to them. But we have seen that it is not.

We have seen that Google is trying to clean up its image, but underneath it remains deeply committed to advancing a progressive definition of the "public good." As its technology continues to become more efficient in command of public debate, it will presumably aggregate even more power and grow stronger than any public authority.

Yes, stronger even than US presidents who can either be co-opted or outlasted—or at least that's the running predicate.

What's Larger—the Amazon, or Amazon?

In terms of ground covered, the sprawling patch of rainforest in South America covers one billion acres and most of six countries. Amazon has a functional reach fourteen times larger, with its presence in 185 of 195 countries on Earth.[45]

In terms of commerce created, there is an estimated $15 billion in revenue generated by enterprises large and small in the Amazon rainforest.[46] Amazon has revenues approaching $200 billion, or thirteen times larger.

- **Amazon's Fortune 500 Ranking: 8**
- **Revenues: $177.9 Billion**
- **Employees: 66,000**
- **Total return to shareholders: 28.9%[47]**
- **Twice as many US households have an Amazon Prime account as have a gun[48]**
- **Half of all online sales made at Amazon[49]**
- **Fast swallowing the retail world**

In terms of resources plowed under, the rainforest has lost 20 percent of lush verdant jungle landscape in fifty years. In half that time, the Amazon juggernaut has been the primary steamroller flattening or plowing under the entire retail world, including major names like Aéropostale, American Eagle, Barnes & Noble, Carrefour, Coach, Finish Line, Ikea, JCPenney, Kroger, Macy's Nordstrom, Office Depot, Sports Authority, Target, Tesco, The Gap, Tiffany, Walgreens and Williams-Sonoma, to name just the better-known casualties that were forced into major downsizing.

An estimated five million species of plants, animals, and insects live in the Amazon, and we are losing 137 every day to deforestation. Meanwhile, Amazon sells over 480 million products and is expanding by 485,000 new products every day.[50]

You might think this is a rather oddball comparison we're drawing here. But it is an undeniable truth that the great Amazon rainforest is receding into the past and another Amazon is rising up to become Earth's biggest store. Just as founder Jeff Bezos intended when he first invented it. Along the way, Mr. Bezos has occupied and laid waste to trillions of dollars of jobs and community merchants along with the gung-ho spirit of an America, now fading away.

It has not been an entirely hostile occupation—not for customers on the receiving end. In its own aspirational twist on the Google and Facebook models, Amazon brought something new to this world that was fascinating and fun: the ability to get almost anything delivered to our doorstep in a day or two, and cheap.

Bezos's seminal insight was that folks enjoy shopping for Chevy Corvettes or Manolo Blahnik lace pumps, not so much for toilet paper or toothbrushes. So, when Amazon made it easier to get these items than it would be running down to the corner store, America—the world—was hooked. Next Bezos solved the problem of the "last mile." He figured out how to get that toilet paper delivered to your doorstep for no extra cost (that you are aware of). Magic!

Which is why, as we write this, Bezos is the wealthiest person in the world. Bill Gates and Warren Buffett are up there with him, but neither sits on top of a company growing at an epic 39.4 percent annual rate over sixteen years—felling one multibillion-dollar sector after another, like so many wounded, bewildered prey.[51]

Bezos Sees the Future: A Quarter of America on the GMI

Like Vannevar Bush, who foresaw how science and not mere bombs would win future wars, or Marshall McLuhan, who envisioned a new age of electronic media that would create a dysfunctional global village, or Richard Feynman who would shrink the world down through nanotechnology, or Tim Berners-Lee who would create the protocols for the World Wide Web, Jeff Bezos has seen the future of retail with an orderly and efficient army of robots fulfilling every commercial need of Amazon Prime customers.

Sixty-four percent of American households are already using Amazon Prime.

In Bezos's vision, there is not much call for a human workforce. He sees the messy hiring and firing of people as an appendage of 20th century morality. So what will happen to the millions of people who can't find work while robots slave away uncomplaining for about a dollar an hour in cost—doing most of the jobs humans once did? Oh, and doing them better, usually?

Again, Bezos has an answer: Guaranteed Minimum Income (GMI). Put a paycheck in the pocket of every American, big enough to cover their basic cost of living, no need for them to work—even if they could find a job. Bezos has seen the future workplace, and there aren't a lot of people drawing paychecks. Robots, cobots, and plain old bots will be taking over most jobs. So that's his vision—robots and GMI…

Bezos sees this as the epic battle of the 21st century—the Meat vs. the Machines—and Bezos is putting his incredible mind and bank account behind the machines. He's basically saying:

"Me, personally, I'm going to become the even-richer richest man in the world and find clever new ways to avoid paying the taxes that'll be needed to give the seventy-three million Americans that the McKinsey Institute estimates will lose their jobs to robots within a decade.[52] If the GMI is set at, say, $25,000…a smidge under two trillion dollars a year should just about cover the costs of this Jim Dandy program."

Of course, Amazon alone won't be responsible for all of the jobs lost to automation. But sighting down Amazon's current trajectory, it will be the biggest contributor. A fair analysis was done by Galloway in *The Four*. He paints a very clear picture of the immediate pain Amazon will deliver:

"If you take the number of people Amazon needs to do one million dollars in revenue vs. the number of people Macy's would need, as Macy's is a decent proxy for retail productivity across the sector (it is, in fact, more productive than most retailers), then it's reasonable to say that Amazon's growth will result in the destruction of 76,000 retail jobs this year. Imagine filling up the largest stadium in the NFL (Cowboy Stadium) with merchandisers, cashiers, sales associates, e-commerce managers, security guards and letting them know that, courtesy of Amazon, their services are no longer needed. Then, be sure to reserve Cowboy Stadium and Madison Square Garden next year, as it's only going to get worse (or better, if you are Amazon shareholders)."[53]

Bezos has been investing heavily in warehouse robotics since 2012, and a tour of one of his 140 fulfillment centers here in the US is a "quiet" experience. There are not many people working the floor in some centers; his goal is to move from not many to none. Once Bezos perfects autonomous vehicles and drone deliveries, the delivery guys will join the warehouse gals sitting at home, waiting for their GMI check to arrive.

And we're supposed to admire the pluck and inventiveness of Bezos's thinking?

Forget about the old "our kids having a better life than us" trope; a nation of nonworking people is a disaster unfolding. That's obvious.

Alexa, Are You Spying on Me?

Any new technology is going to go through some initial growing pains. That's why every responsible company will try a product first among employees and friends in "alpha" tests, and then with early adopters in "beta" tests—in order to catch bugs and patch flaws. Then, hopefully, the number of bugs will have been narrowed to a manageable few, and the product will be ready for general release.

So, when you discover a tech product out in the marketplace, with a full complement of advertising that's driving consumer adoption, and the product is doing wonky and even dangerous things, it's hard to contain your suspicions and mistrust about the product, and about the people behind it.

To wit, Amazon's release of its new virtual assistant, Echo. Amazon unmistakably wants to drive commerce from their website to Echo, at least at the outset. So much so that they'll offer a lower price if you shout a verbal order to Echo than if you click on their website—for the same item. This is their prerogative, but did they release too early?

One Portland, Oregon, family set up an Echo so they could control their lights, HVAC, and home security system using voice commands. Cool, they thought. Until learning that the Echo's operating system, aka Alexa, one day decided to record a private conversation the family was having and then send an audio file of it to a random contact in their contacts file. "I felt invaded," said the homeowner, "I'm never plugging that device in again, because I can't trust it."[54]

This couple is not alone in their frustration over the privacy violation and the damage it could have done. Then there are the security risks that have not been locked down.

A new breed of felon, known as a "voice squatter," has arisen. These are unseen hackers who break into the code of one of the third-party developers on the device and trick it into doing something nefarious. For example, the user says, "Alexa, open Capital One," which is a legitimate banking app on the Amazon device. But the hackers create a twist on the command so that if the user inadvertently varies the request even slightly, such as saying "Alexa,

Capital One" or "Alexa, Capital One, please," the hacker can take control of the device unbeknownst to the user. At that point, the hacker can go on stealing sensitive financial information from the user until the breach is noticed.[55]

A Burger King commercial is cleverly scripted to trip the Echo to describe its Whopper burger. A fun bit of marketing. But not so much when hackers send you an official-looking email that compels you to click a link that plays an audio file that commands Alexa to unlock the door to your smart home. The hackers can even modify the audio frequencies, so you don't hear the command, but your door lock does.

It has happened.

The Good, Bad, and Uncertain of Home Assistants

A device expert will tell you that Amazon's Echo is always listening, but not recording. It is listening for its "wake up" word, such as "Alexa." Then it starts recording and trying to fulfill your request. Yet there are many stories of Alexa's so-called smart speakers waking to the wrong word or misinterpreting fragments of speech for commands to act. And yes, these flaws have caused the devices to start recording—even when not commanded to.

All of your requests of Echo are kept on file as well. All of your voice recordings become part of your record, just as all of your online browsing does. These recordings join your emails, photos, and documents that can be tied to you, by bad actors and good.

Since Amazon and Google want these virtual assistants in every home, along with their other connected devices, the level of privacy and security that we all experience is going to steadily erode, while the potential for mischief steadily builds to sometimes dangerous levels. So what protection do we have?

The best protection is a vibrant market with many innovative companies in fierce competition to overcome the shortfalls of existing products. This kind of innovative capitalism is what built America, until it was lost to a marketplace that has effectively only four players dividing up the spoils of monopoly.

Restoring the opportunity for innovation—that's what policymakers should be looking at.

A Race to World Domination: It's Amazon's to Lose

In the race to become the first trillion-dollar company in valuation, Apple beat out Amazon in 2018 by only a month—and that surprised many analysts. The betting had been on Amazon. In the last two years:

...tock doubled while many competing retail

...oeing 757s to its fleet, along with thousands of
...early 500 Whole Foods stores, giving it intelli-
...tribution points within twenty miles of half of

...e country on Amazon Prime. This drives the
...spend $1,500 a year buying goods through
...s for non-Prime buyers.[57]

...e with a single click for two-day, one-day, even
...d zero-click ordering is in testing; with this you
...your home, so they can put the goods right into
...igerator, incidentally driving up your purchase

...illions of products available, a SKU to meet every

...is on call 24/7, taking orders in upscale homes.
...g is: Consumers first do their research on brand
...k" over to Amazon to make the purchase—since
...mparable, shipping usually free. But several retail
...that 55 percent of all product searches now begin
...t's a steadily growing number.[58]

...r 50 percent of online sales in the US and is growing
...tegories where it had previously experienced head-
...ds and food.

...018, when Amazon doubled down on discounts,
...n one hundred million items, including one million
...s, raking in an estimated $2 billion in sales.

...ingness, Amazon owns the world's most trusted
...y not just kick back and let Bezos run the world?
...osition to run the house on retail. He could scoop up
..., Sears for $135 billion, Target for $45 billion, Macy's
...dstrom as well, for giggles. He could even offer a 10
...h of their shareholders—to seal a quick deal. After
...ould only incur a 60 percent dilution—something

Advance Praise for
Big Tech Tyrants

"Most Americans have no idea of how much information Google, Amazon, and Facebook have gathered on them. *Big Tech Tyrants* is your reference guide into the world of big data surveillance without the jargon. It will help you understand why privacy matters."

—WAYNE ALLYN ROOT, RADIO AND TV PERSONALITY

"Floyd Brown and Todd Cefaratti as digital publishers have had a front row seat to watch the changes in technology that have revolutionized: news, politics, and commerce. They will provide you insights into how Big Tech Tyrants use data to manipulate how and what you think. These companies make you dance to their tune."

—MIKE HUCKABEE, FORMER GOVERNOR AND HOST OF *HUCKABEE*

"I have spent my career in business, politics, and media. These three sectors of our economy have been whipsawed by technological changes. *Big Tech Tyrants* is the most comprehensive exposé I have seen about the dark side of the forces of big data collection and the misuse of it by large tech firms."

—HERMAN CAIN, FORMER CEO AND MEDIA COMMENTATOR

"Brown and Cefaratti provide brilliant insight as to how the collection of personal data is daily misused and exploited by the modern-day robber barons of Silicon Valley. Until you read *Big Tech Tyrants*, you won't understand how much data these firms collect and how they use it to manipulate your very reality."

—DAVID BOSSIE, DEPUTY CAMPAIGN MANAGER, TRUMP FOR PRESIDENT

Bezos and his shareholders could manage—in exchange for controlling more than half of the entire retail ecosystem, free to treat the country like his own company town.

Of course, Bezos would not be likely to scoop up all these competitors—surely someone would step in, right? Well, nobody has lifted a finger to stop him yet!

Or perhaps you say, as many do, that sure—Amazon takes 50 percent of online sales but only 5 percent of all retail sales—so warnings of an imminent takeover are grossly exaggerated.

It's true. Amazon was on track for 5 percent of online retail sales by the end of 2018.[60]

But it's not Amazon's relative market share that should disquiet us. It's the way in which Bezos's business model has weakened the businesses of the legacy retailers. They have become brittle at the core, and vulnerable. That's why so many have fallen or are now struggling to stand up. They've been hollowed out, as if by termites, and you can't see how much structural damage has been done.

But it has been done.

If the hollowing out continues long enough, and one by one the major retailers crumble, Amazon won't need to buy them. It can step in and replace them and watch as its market share increases rapidly and a monopoly locks in, unassailable in the midst of the wreckage it caused.

Our past several Justice Departments didn't bat an eye at the plainly visible concentration of the retail sector. Hopefully the current Justice Department will see the trajectory we're on and take steps to make the American economy more competitive, not less.

> **Facebook and Google own online media. Apple owns the phone. Amazon is positioning to own retail.**

Amazon Is the Original "Frenemy"

Amazon has developed a three-step commerce killer:

1. **Cripple local merchants by undercutting them on price.**

2. **Offer to "befriend" and save those merchants by allowing them to sell on the Amazon platform.**

3. If those merchants then become successful on the platform, Amazon creates their own brand version of the product and promotes it at the head of the listings, where 75 percent of purchases are made.[61] (As the saying goes, the best place to hide a dead body is on page two of the search results.)

So, Amazon is playing a form of three-card monte where the merchant can actually follow the face card, but still loses.

From His Garage to Rivaling the British East India Company

As the world's richest man, Jeff Bezos now lives in the constant glare of both public admiration and condemnation. He has fascinated us all with his accomplishments, while at the same time creating so many structural problems in the economy that any balanced analysis of the man contains both praise and scorn.

Knowing as much himself, Bezos has done what most monopolists do— begun making high-profile donations to charity and, when appearing in public forums, making sure to tell the story of his humble beginnings. And his is a tremendous story.

Bezos did launch Amazon out of his garage, and his success was never certain. He made a lot of big mistakes along the way. So, we have nothing but admiration for what he has accomplished. Our concern is the price our nation is increasingly paying for it all. This is because Bezos has figured out how to control the entire retail supply chain—not only online, but out in the brick and mortar world, as well.

Gaining that control begins with groceries.

Selling groceries well is the key to selling everything else well. That's because groceries are the hardest thing to sell at retail. Each banana, for example, is unique, and bruises easily, and quickly goes from green to ripe to mush. But twenty cents of every dollar we spend on retail goods goes to groceries. That's about $800 billion a year. And it's money spent every week, reliably. So, if a company can get groceries right, they can get into a family's regular purchasing stream. That's what Amazon is trying to do, though now a small percentage of groceries (about 4 percent) is purchased online.[62]

For Bezos, owning our digital spending is not enough. By attempting to seamlessly link both realms, Amazon is positioning to be part of every single purchase we make. That's why Bezos bought Whole Foods—to further ingrain

Amazon in our routines and behaviors and complete his takeover of the retail supply chain. By coupling...

1. our purchases of perishable items with
2. our existing purchases of hard goods, many of which are made by Amazon, and
3. a fast-moving delivery system like Amazon Fresh and Prime Now

...Amazon gains control over the complete retail supply chain from farm to fridge, and from raw materials to the recycle bin, as they say in the trades.

This kind of control by a single company will go the furthest to match the reach of history's most powerful company, the British East India Company, which beginning in the 1600s controlled global commerce, ruled large swaths of land, and even fought its own wars while essentially holding the British government under its thumb.

Yes, that's where Amazon is headed. And we've only touched on part of the plan. Think of what it will mean when Amazon rolls its business model across one industry after another.

Other Industries in Line to Get "Amazoned"

In addition to dominating the $3.53 trillion retail sector, Amazon is positioning to control other sectors of the US in a practice that has been allowed to go so long unchecked that it even has a nickname—getting "Amazoned." It requires no definition—the scorched-earth destruction of industries in Amazon's sights is so similar to the tractors plowing up the South American rainforest that the visuals cement easily in mind.

In all, Amazon is aiming to control industries now generating $5.37 trillion in annual revenues, or almost 30 percent of the nation's GDP. Now standing in front of the Amazon company tractors are...

Financial services—a $1.45 trillion industry

Taking over the entire banking and insurance industries sounds like a challenge even for Amazon. But it is moving to bite off a large chunk. By first providing checking accounts to millennials, Amazon gets a toehold and grows from there. Analysts at Bain estimate that within five years Amazon could have seventy million banking customers.[63]

Drugstores—a $272 billion industry

Amazon recently acquired PillPack, a mail-order drug-seller valued at $361 million, to give the company a national foothold in the prescription-drug market.[64] PillPack could also be a useful tool for Amazon's mysterious partnership with Berkshire Hathaway and JPMorgan Chase to disrupt health care delivery in America. We have no idea what may come from this acquisition or the partnership, except further concentrating and pricing power for Amazon.

Delivery—a $100 billion industry

Amazon is pitting its logistics prowess against UPS, FedEx, and USPS. In a Los Angeles test market, the "Shipping With Amazon" pilot project is proving that Amazon can come out on top in this sector as well.

Event Tickets—a $19 billion industry

Market leader Ticketmaster doesn't get much consumer love, and Amazon knows this is a sector stuck in the past and ripe for disruption. And ideal for disruption because it is estimated to grow from a $19 billion to $68 billion industry by 2025.[65] Amazon has run a ticketing business in England and is reported to be preparing to roll into the US to capture this growing industry.

Amazon is also moving into the clothing and furniture business, intending to use its virtual reality tools to make online shopping for these items appealing and a lot more convenient than actually visiting stores and showrooms.

With each new industry that Amazon moves into, it will usually improve or enhance the experience for the end consumer. Few can argue that. But when so much of the American economy is controlled by one man, the larger interests of society are at risk.

Bezos Extends His Power Over Washington

Carefully micro-managing his ascendency, Bezos in 2016 bought the former Textile Museum, creating the biggest private house in Washington, D.C. He also bought an adjacent property. At 27,000 square feet when the renovation is complete, the combined mansion will be 133 percent the size of the president's living quarters. A message Bezos no doubt intended to send.[66]

Bezos aims to throw the biggest bashes in town in his palace, with its twenty-five bathrooms, eleven bedrooms, five living rooms, two kitchens, two libraries, two exercise facilities, two elevators, and a resplendent ballroom with floor-to-ceiling Ionic fluted columns, limestone fireplace, and balconied promenade.

Imagine his notes to the architectural team: "Be sure it dwarfs everything in town, especially the President's living quarters."

We expect that Bezos plans to locate Amazon's headquarters in nearby Virginia, in order to exert as much control over the government as possible.

This is the battle he is waging with Google, a formidable adversary. A true "King of the (Capitol) Hill" battle. Each platform can choose the politicians they wish to carry water for them. Plus, Bezos, owner of the *Washington Post*, has one of the world's most influential newspapers in his pocket. He is reportedly a hands-off owner—which he can continue doing as long as the newspaper's leftist slant mirrors his own.

Lastly, Bezos has armfuls of federal contracts, as well as the intelligence community's confidential files, in Amazon's cloud (Amazon Web Services). By almost any measure, Bezos is on track to become the most powerful man in a town full of them. If allowed.

STRIKE 3

Thought Manipulation

The Fastest Descent of Any Industry in History

For most of history, the easiest way for despots and other authoritarians to block the spread of an idea was simply to silence the source. Shutter the newspaper. Strong-arm the TV exec. Officially censor the publisher. Hold a loaded gun to the announcer's head. It was quite effective, brutal, and corrupt.

We fought to create a world free of these abuses.

And we thought we had truly entered a new and enlightened time, a truly connected and globally networked world where all the old oppressors didn't stand a chance—because in this new world, no idea could ever again be censored.

Even the smallest voice could blossom and live forever in this global town square.

A golden new age of free speech, with all of the barriers to entry ripped away, the public commons open to anyone with a connected device and opposable thumbs.

Wonderful thoughts, enlightened thoughts. Yet, we would ultimately learn, naive thoughts.

While we marveled at one Silicon Valley accomplishment after another, something else was going on. The subjects of our marvel were becoming our gatekeepers. Pay no attention to the wizards behind your screens. Embrace the wonder, they told us. Extol the technology that will give you tools to transform this world on unprecedented levels. And while we watched, and embraced, and extolled, the creators locked our doors and became our gatekeepers.

We're speaking of the youthful geniuses behind Facebook, Google, and Twitter, primarily.

Young, brash, and eager to break things, these new Boy Kings (as Facebook employee #51, Katherine Losse called them[1]) were soon deciding what up to 90 percent of the world sees or doesn't see online. The Boy Kings and their social platforms. Controlling the diffusion of data, as they call it. Deciding what people will read, see, and hear, and never knowing or caring what anyone else is reading, seeing, and hearing, because the social platforms give everyone their own feed (content)—no two are alike, so no one has any idea what anyone else is seeing. And to keep the people hooked on the platforms, to keep them returning, the cleverest addiction algorithms were then deployed, so that people would find it hard to leave—like giving up cigarettes. And after all this, the Boy Kings would begin to believe they could exert their own political views over the populace.

Boy Kings, or Big Tech Tyrants, as we call them, give a powerful visual. And if they are accurate labels, it suggests that these social platform operators have gained a non-elected (or, should we say, a self-elected) authoritarian power and wield it with more effectiveness and precision than any sitting government or military strategist.

They certainly know more about the citizens of America (and many other countries) than our own government does. And knowing the people better than the government, in this modern age of data, means having more control over them. Certainly more control than at any time since John D. Rockefeller and Standard Oil lorded over America. These Big Tech Tyrants were nurtured on the teat of the programmer's classic conditional "if-then" statement: *if* people are doing this, *then* we can manipulate them into doing more of that, or…something else.

It's that will to manipulate that we'll look at now.

The Eleven-Year ZAP (Zuckerberg Apology Tour)

To those following politics, 2018 was yet another year marked by social platform executives trooping up to Capitol Hill to explain themselves to Congress.

In May 2018, Mark Zuckerberg faced off with the House Judiciary Committee, doing his best to swat down questions from the lawmakers, who, for their part, didn't appear to know a meme from a memo, an algorithm from an avatar.

The following month, Zuckerberg would be in Brussels, testifying before European Parliament lawmakers who had their own list of concerns about the abusive social platforms.

Both in Brussels and in Washington, Zuckerberg played his role brilliantly—striking the pose of a contrite and apologetic technologist, admitting, "It was my mistake, and I'm sorry", when examples of conservatives like Diamond and Silk were suppressed on their platform for no valid reason. It was a line he had become very accomplished at delivering....

ZAP 2007: Revealing Very "Personal" Data without Permission

Zuckerberg's long history of breaking things and apologizing later, of dazzling and then fast-dancing his way out of problems, goes way back to November 2007—just three years after he dropped out of Harvard. That's when Zuckerberg's special strain of privacy hostility first played out publicly with a new Facebook app called Beacon.

With Beacon, a Facebook user could be alerted anytime a friend was purchasing something somewhere. It was billed as a community-building feature; it was also the first big violation of user privacy in the then-brief history of the internet. And it was offered to Facebook users on an opt-out basis—meaning Facebook automatically sent out these alerts every time—unless you told it not to every time. Which was, by design, a non-trivial, time-wasting, frustrating process. So if you were any kind of Facebook user, you would find yourself...

- Stuck in a stream of constant notifications.
- Trying to turn them off.
- Choosing to deal with them.
- Getting angry.

The new Beacon feature was, by every assessment, an epic fail from the get-go. But Zuckerberg, a brilliant engineer known to have the social awareness of an armadillo, thought he knew what his users wanted. So he kept Beacon alive, even as the complaints piled up in Menlo Park. Weeks passed; finally, there was a post on his blog: "We've made a lot of mistakes building this feature, but we've made even more with how we've handled them."[2]

With that, Zuckerberg wound down Beacon ignominiously and settled a $9.5 million class action suit against Facebook. If the Boy King learned anything from the catastrophe, it would not soon be apparent.

ZAP 2011: Arbitrarily Overriding Users' Indicated Preferences

Fresh off the Beacon debacle, Facebook settled on a new way to manipulate its users.

With no warning, on what was probably a sunny day in the Valley, Facebook flipped a bunch of their users' settings from "private" to "public by default"—just to see what would happen.

Just to see what would happen.

With not even so much as a "How do you do?" all the data users handed over to Facebook, and had secured with confidence by clicking on the privacy box, was suddenly going to be made public.

Privacy was pushed over a precipice that had once been guarded by a tall security fence, leaving exposed for all to see not just your name or gender, or all your photos, but also your entire list of friends as well as potentially more sensitive information—sexual orientation, business dealings, and political views, along with plenty of "who cares" information.

With the flip of a switch (yes, overly simplified, but just as fast), it all became public information. All kinds of wheels began turning, and not all of them in a positive direction.

In November 2011, the FTC stepped in and charged Facebook with "unfair and deceptive" behavior that threatened the health and safety of its users. Once again, Zuckerberg issued his most heartfelt apology, agreeing to a twenty-year probationary settlement. Under the terms, Facebook would submit to independent audits of its privacy practices every two years. Any violations would result in fines of $16,000 per day, per violation.

This rebuke to Facebook was meant to turn the page on the whole fiasco. And Zuckerberg dusted off an old blog, writing with all the heartfelt candor he could muster: "A small number of high-profile mistakes" had been made, and Facebook was committed to upholding better privacy controls going forward.[3]

But dollars to doughnuts, he had never given the matter more than five minutes' thought. Privacy had never been in his lexicon, and certainly not in the business model that made him rich.

Because sure enough, as day follows night, Facebook would again be pushing the bounds of the settlement with the FTC and intruding on its users' privacy in even more inventive ways. As we'll see, Facebook's violations of the terms of that 2011 settlement should, in fact, result in some rather mind-boggling fines.[4] If we've got the math right:

Facebook's Big Whopping Fine
If found in violation of the FTC's 2011 consent decree:
$40,000 maximum fine per day
x estimated 20 million users affected
x estimated 90 days in violation
= $72,000,000,000,000

(72 trillion, or four times the US GDP)The FTC is currently investigating. However, our own admittedly imperfect reading of the consent decree, including the allowance for up to $40,000-apop fines for Facebook, suggests the social platform could realistically be looking at a government fine of $72 trillion, or four times the 2018 US gross domestic product!

We'll see.

ZAP 2012: Toying With Users' Mental States

It was an election year. Some Cornell data scientists joined with Facebook engineers who wanted to scratch an itch:

How far could they go in manipulating their users?

1. Could a single user's emotions be toyed with in such a way that it might affect all that user's friends?

2. Could it spread the way a contagion spreads, with the emotions of many thousands—or even millions—of people being changed simply by what was or was not placed in Facebook's News Feeds?

3. Could Facebook trigger a tsunami of passionate support for an individual candidate or issue, for instance?

And so, for one week, in the manner scientists play with lab rats (and unbeknownst to nearly 700,000 randomly selected rats…er…users), Facebook played with News Feeds. In the experiments, Facebook dialed up the number of "positive posts" that one group would receive and dialed down the number of "negative posts" another group would receive. Publicly, Facebook insisted:

We did this research because we care about the emotional impact of Facebook and the people that use our product...We felt that it was important to investigate the common worry that seeing friends post positive content leads to people feeling negative or left out. At the same time, we were concerned that exposure to friends' negativity might lead people to avoid visiting Facebook.[5]

And perhaps that was also a concern the researchers sought answers to. But at the core of their tinkering was a fundamental question:

Could Facebook play with the content that users saw and actually modify their emotional state as a result?

Yes; yes, they could, they discovered. Turns out, users who had "positive posts" removed from their News Feeds made fewer positive posts and more negative posts of their own, and vice versa. The explanations for this user behavior ranged from "it's an online version of monkey see, monkey do" to "it's like keeping up with the Joneses."

As for the Cornell data scientists, they published their findings in the *Proceedings of the National Academy of Science* in 2014 and were proud of having produced "the first experimental evidence" that a "massive-scale emotional contagion" could be launched on the social platforms.

(Dear reader, pause here for a moment and reread our Forewarning. We are reluctant to believe in coincidences, when they align oh so neatly.)

How surprised the Cornell data scientists must have been, then, when Facebook users went ballistic. Hardly anybody thought Facebook was even *allowed* to run such experiments, much less that it would be eager to do so. Two in three Facebook users have no idea that when they tick the AGREE box at sign-up that they are authorizing Facebook to "use the information we receive about you...for internal operations, including troubleshooting, data analysis, testing, research and service improvement." So, yes, Facebook was legally entitled to treat its users like lab rats. But ethically? Doing so without even telling them?

Comments exploded on Twitter:

"Does everyone who works at Facebook just have the 'this is creepy as hell' part of their brain missing?"

—@SARAHJEONG

"Impressive achievement by Facebook to snatch back the title of most dystopian nightmarish tech company."

—@TOMGARA

In the weeks it took for this new scandal to blow over, it was natural to assume that Facebook would lick its wounds and institute yet another new policy forbidding the experimentation on people without their consent.

But that was far from the case.

In her book, *Who Can You Trust? How Technology Brought Us Together and Why It Might Drive Us Apart*, Rachel Botsman tells the much more comprehensive story of Facebook's testing programs. It's a story that few know. As a result of her interviews with Facebook data scientist Dan Ferrell, she revealed: "At any given time, any given Facebook user will be part of ten experiments the company happens to be conducting."[6]

Botsman concluded that the statistical likelihood that you have been a lab rat in one of its experiments is 100 percent. Even the single revelation of Facebook's lab-ratting had struck a deep nerve with people. Imagine how much more anger would follow if it was widely known that the testing is ongoing? It would, no doubt, rightly be perceived as a scorching betrayal of trust. But as troubling as this betrayal was, it also raised a far more profound question:

If Facebook can manipulate a person's moods with an unseen tweak of its algorithms, what else could the social platform control? Could it control who we vote for?

ZAP 2016: Facebook Jumps the Privacy Shark

2016 was a busy year for Facebook. Within the course of a few months, America learned…

1. **Facebook's pages were deeply infested with Russian trolls.**

2. **Facebook gave troves of data to the Trump campaign.**

3. **Facebook had cornered the news market and now controlled it.**

4. **Facebook was censoring conservatives (though the world didn't know it prior to 2016).**

It was as if Facebook had lost its moorings, run out of good ideas for the platform, and adopted its own version of Fonzie's "Jump the Shark," attempting one more desperate stunt off the ski ramp of social media, trying anything to get a rise out of viewers.

And with one Facebook effort after another exploding in their faces, they swung into damage-control mode, initially insisting that none of this really happened, or if it did, it wasn't true. It was the kind of logic that barely raised an eyebrow in a "fake news" all-the-time climate. But it was not the kind of logic that would hold up long to scrutiny.

First up in 2016, some preliminary reports surfaced about Russian hackers, bots, trolls etc. causing mischief on social platforms, primarily Facebook. Though, technically, these same hackers had been doing this very thing worldwide, especially in the Ukraine since 2014, and it was all well-known to anyone with a Level 1 clearance in Washington.[7] These Russian hackers, or trolls—because it's a sexier way to demonize an enemy—were doing it by the book. Literally. The Russian book titled *Information Psychological War Operations: A Short Encyclopedia and Reference Guide.*

This 495-page encyclopedia was standard issue for Russian's top political technologists, state security services, and civil servants. But for the excellent sleuthing of the Soviet-born British journalist, Peter Pomerantsev,[8] it is not likely anybody in the West would have ever heard of this book. (Pomerantsev is also the author of *Nothing Is True and Everything Is Possible: The Surreal Heart of the New Russia,* a fascinating read on a crumbled post-communist nation.[9]) Pomerantsev told how Vladimir Putin escalated the information warfare game by mobilizing armies of paid trolls to create fake news stories complete with doctored photographs and video clips, basically weaponizing data in order to sow discord in free nations.

Disinformation was targeted "like an invisible radiation" upon opponents so "the population doesn't even feel it is being acted upon" and the government "doesn't switch on its self-defense mechanisms."

That a malicious disinformation campaign was being waged was hardly surprising to national security operatives in Washington. Among US intelligence agencies, there is long-held agreement that Russia has been successful at interfering in elections around the globe. It is also true that the US engages in similar attempts. But the results of this particular Russian adventurism in 2016 was believed to have a far greater impact on our election outcomes—thanks to the design of the social platforms, especially Facebook.

Russian operatives ran 3,000 ads on Facebook—a very small number compared to the total count across the campaigns. But those ads favored candidate Trump over candidate Clinton…which gave the leftists a hall pass to throw an outrage party. The left mobilized their outrage warriors and threw their liberal activist army into hyperdrive.

To this day, experts disagree on just how effective the ads were. The cost of the ads was reportedly $11,000, while the official Clinton and Trump campaigns alone spent $81 million on Facebook ads.[10] That means only one of every 25,000 items that typical Facebook users saw in their News Feeds came from a Russian. And since the content was all but indistinguishable from the authentic ads smeared all over Facebook, perhaps the Russian influence on the American voters could not have been much more than one marble in a jar of 25,000 marbles.

Other analysts insist that while the Russian ads may have been few, their total number of "shares" was outsized. Those few Russian ads were apparently shared 340 million times quantifiably, billions of times potentially.[11] They were shared so crazy fast because their writers had been crafty enough to mimic the kind of hornswoggle that Facebook users just love to share out to possibly hundreds of millions of screens.

When news first broke of the Russian campaign, Facebook leapt into damage control mode once again. Their security researchers released a big white paper very publicly and commented with appropriate head-hanging solemnity, if only to hide the crocodile tears:

> We have had to expand our security focus from traditional abusive behavior, such as account hacking, malware, spam and financial scams, to include more subtle and insidious forms of misuse, including attempts to manipulate civic discourse and deceive people.[12]

Facebook wanted the world to know that it cared, that it was going to get out ahead of this insidious contagion of dark ads and fake posts perverting their platform. But Facebook's problems were less about what nefarious bad actors were spreading across the platform, and more about what Facebook knowingly let happen. Facebook took in an estimated $1.4 billion in digital ad spending during the 2016 election.[13]

A lot of that cash came from the Trump campaign, igniting another furor.

The Trump campaign had retained a then-unknown firm named Cambridge Analytica. Their secret weapon was Aleksandr Kogan, a Cambridge

University academic who had crafted a "quiz app" that coincidentally harvested a trove of Facebook user data. He then sold that data to Cambridge Analytica, which, in turn, used it for its client, candidate Trump.

Cambridge Analytica had access to the details of fifty million Facebook users, without their knowledge or express consent, and who were completely unaware that in the fine print of the quiz app, there was a line about the user agreeing to allow their data to be used for commercial purposes. So this quiz app was, by Facebook's own policy book, an outright violation of the company's rules. Yet not once in two years after the quiz app ran did Facebook say anything. It didn't appear that Facebook ever checked to make sure that apps were complying with its rules.

Kogan made that very argument and, feeling wronged, he made it to anyone who would listen, laying the blame at Facebook's feet. He did nothing wrong, he insisted, because Facebook never actively enforced its policies and thousands, if not millions, of other developers were doing the same thing he had done, and had done so regularly.

In fact, Cambridge Analytica *was* just one of many outside companies harvesting Facebook data in 2016. And Facebook was well aware of it—knowing that it was good for business—and said not a word about it until outside parties shined some sunlight on things.

Facebook was essentially handing over its user data to outside companies, even while insisting loudly over and over that they do no such thing.

The 2011 FTC consent decree Facebook signed forbade the company from releasing users' data without users' consent. Should the company do so, their action would trigger a massive fine.

Soon the proof of Facebook's violations would surface. Investigations would find that Facebook had been sharing users' data—without their consent—with fifty-two hardware and software makers and sixty-one app developers, including the Chinese company Huawei, which the US government had deemed a threat to national security.

Also allowed to hoover the Facebook database (and snatch up information on users' friends such as name, gender, birthdate, current city or hometown, photos, and page likes) were big names like Apple, Amazon, Microsoft, Samsung, Alibaba, and Qualcomm.

Joining in the data orgy were other mobile carriers, software makers, security firms, and chip designers, as well as thousands of lesser-known dating apps, chat apps, games apps, music streaming apps, data analytics apps, and news aggregator apps.

The revelations of the sheer number and reach of these privacy breaches left everyone from Republicans on Capitol Hill to progressives at the ACLU wondering about Facebook's (man)handling of user data and whether or not it really "locked down the platform" back in 2015 when it claimed to have done so.[14]

Once again, Zuckerberg was forced to issue a big apology, and he agreed to end forty-five of the 113 data sharing partnerships as soon as doable. Of course, Zuckerberg did not admit to violating the 2011 settlement with the FTC forbidding the release of user data without permission. His lawyers must have advised him to say that, since clearly he had run deep into violation territory. The FTC has a file open on the matter and will make its own determination...which could include up to a $72 trillion fine, or whatever settled fraction of that fine is acceptable to a government agency doing its job.

As 2016 continued to unravel for Facebook, even the most casual social media user began to realize:

1. **The bulk of the news people were getting was coming through the Facebook and Google platforms.**

2. **This news delivery process was not only hidden from view, it was acting strange.**

3. **The problem was not only in the "hands off" approach that Facebook was taking, but in the "hands on" approach.**

In May, several politically conservative news publishers began to see strange things happening to their Facebook audiences. At first, none of these conservatives had any idea what was happening. It was just too unbelievable to imagine that Facebook would be actively toying with the news they routed into users' News Feeds.

With all the scandals and potential punishments that Facebook had been through and was still facing, why would they risk further upsetting the powers that be in Washington, D.C.?

Hubris, we would soon learn.

Facebook employees had come to a point in their personal development where they realized that *they* knew the truth in the news, and the quality in news organizations, and that's that.

What's more, they no longer believed that simple Facebook users could make those determinations about truth and quality on their own.

This is not our own conjecture.

This is Campbell Brown, former anchor at NBC and CNN and now head of Facebook's news partnerships team, telling attendees at a recent technology and publishing conference that Facebook would be censoring news publishers based on its own internal biases:

> This is not us stepping back from news. This is us changing our relationship with publishers and emphasizing something that Facebook has never done before: It's having a point of view, and it's leaning into quality news…We are, for the first time in the history of Facebook, taking a step to try to define what "quality news" looks like and give that a boost.[15]

Hubris writ large.

The engineers of Facebook, adept at building a technology platform, were now becoming…journalists at a news organization?

Even the *New York Times* was wondering about this turn. They began running in-depth features about this new power that Facebook had accumulated and appeared willing to exercise. In a wonderfully headlined feature, "Inside Facebook's (Totally Insane, Unintentionally Gigantic, Hyperpartisan) Political-Media Machine," writer John Herrman peeled back the Facebook operation and showed how the social platform had "centralized online news consumption in an unprecedented way."[16]

Over at *The Atlantic*, Alexis C. Madrigal did an in-depth feature on Facebook's powerfully non-neutral force on electoral politics. His story, titled "What Facebook Did to American Democracy," concluded that "Damn, Facebook owns us."[17]

ZAP 2018: Year Eleven of the Unapologetic Apology Tour

Another year found Zuckerberg and assorted Facebook execs back on Capitol Hill on the latest "break things then apologize" tour and charm offensive. Though by now the list of "broken things" had grown so large, and so well known, that Mr. Zuckerberg must have felt like he was testifying for his company's life. In fact, he was; Senator Dianne Feinstein made that clear: "You've created these platforms, and now they're being misused, and you have to be the ones to do something about it…Or we will."[18]

In the hearings, the American people learned many things. Perhaps most glaring, we learned that Zuckerberg never planned to tell us about any of this. Zuckerberg knew back in 2015 that Cambridge Analytica had scooped up all that data. He told them to stop. Then he forgot about the matter, apparently— or just conveniently. At the time, Facebook could have notified the estimated 87 million people affected by Cambridge Analytica. It would have taken them five minutes to do. But it didn't.

"In retrospect, that was clearly a mistake," Mr. Zuckerberg admitted. Well, duh!

In one set of hearings, Facebook's head of global policy management, Monika Bickert, was grilled by Rep. Matt Gaetz (R) of Florida.[19] The congressman wanted to know why, if Facebook was truly trying to protect its users, did the social platform allow a page called "Milkshakes Against the Republican Party" to remain up and active for months, even while it repeatedly called for violence against Republicans. And then the tragedy finally happened when Bernie Sanders supporter James Hodkinson attacked Republican members of Congress at a practice, with Steve Scalise being shot and almost losing his life: "How many times does a page have to encourage violence against Republican members of Congress at baseball practice before you ban the page?"

Bickert's staff had already responded to a written complaint by Gaetz's staff, saying Facebook didn't have any problem with the Milkshakes people when they published this dark and cracked gem: "Dear crazed shooters, the GOP has frequent baseball practice. You really wanna be remembered, that's how you do it."

Facebook's official position? The group's page "doesn't go against our community standards."

But since advocating for murder is clearly not protected speech under the Constitution, Gaetz pressed Bickert for some insight into Facebook's community standards: "How many times does this have to happen before the page comes down?"

Bickert replied: "When a certain threshold has been met," and that "threshold varies depending on the severity of different types of violations," and thus varies from case to case.

Though she may not have known it, Facebook's policy chief was making the case that Facebook was, in fact, exercising a particular bias in their content delivery systems.

It's a bias that's not easy to quantify—and for a lot of reasons.

Some would insist that engineers like Zuckerberg can just sit up in their silicon towers and throw the switches to make anything they fancy come true. But of course this is just fantasy, and the problem actually runs much deeper. Facebook reportedly has more than a billion lines of code sitting beneath the shiny surface of their website. So much code, and much of it written by the chaos monkeys we've referred to, that nobody really knows what's really happening all the time. As Christopher Mims, the tech writer for the *Wall Street Journal*, has pointed out: "The inner workings of Facebook's data-harvesting behemoth are so byzantine, that in some ways Mr. Zuckerberg is just as confused as the rest of us about how it all works."[20]

So if the guy who is known to run his company like a benevolent dictator, and who is known to be actively involved in every major decision about its going forward, has lost control of his creation, then quite clearly the creation has grown too big.

For example, in one round of testimony on the Hill, Zuckerberg told a Senate subcommittee hearing what Facebook is doing to protect user privacy: "Everyone should have control over how their information is used" and that if you'd like, "you can get rid of all of it."[21]

But as Mims showed in his investigation, that's just not true. In fact, there are several things about Facebook over which users have no control; we don't get to "opt in" or remove every specific piece. Often, we aren't even informed of their existence—except in the abstract—and we aren't shown how the social network uses this harvested information.

Facebook keeps tabs on lots of things, including your browsing history. This history is stored on a server housed in big, quiet, air-conditioned server farms surrounded by barbed wire. But if you ask Facebook to "download your file," you won't find your browsing history in the data set. So when Zuckerberg says you can control your information and even get rid of it, he actually means you cannot control your information and you cannot get rid of it.

Earlier in the year, before being called yet again to testify on Capitol Hill, Zuckerberg had not sounded worried and felt as if 2018 was going to be another business-as-usual year. His New Year's blog post was almost upbeat:

The world feels anxious and divided, and Facebook has a lot of work to do—whether it's protecting our community from abuse and hate, defending against interference by nation states, or making sure that time spent on Facebook is time well spent…My personal challenge for 2018 is to focus on fixing these important issues.[22]

But a few months later in the hot seat on Capitol Hill, he had changed his tune. He was suddenly eager to find a way to keep his sprawling mess of a company from being taken apart. There is an old saying among combat veterans: "There is nothing quite like being shot at, and missed, to focus your attention on the problem." Wishing to avoid another—possibly fatal—shot at the hearing, Zuckerberg testified: "I think the real question, as the Internet becomes more important in people's lives, is what is the right regulation, not whether there should be or not."[23]

He sounded like a man suing for peace. He could (finally?) see that he'd crossed too many lines too many times. So he'd better try and have a voice in the crafting of his surrender from private company to…whatever the future might hold for him.

Following the dressings-down by legislators, Facebook began making a big public show of fixing its laundry list of problems. It began cutting ties to third-party data brokers, rewriting its terms of service, and running audits on developers who can access Facebook user data.[24]

Facebook's head of News Feed, John Hegeman, got out in front of the PR offensive: "We're doing everything we can to fight this. 99 percent isn't good enough."[25]

On a wider front, Facebook's head of product, Adam Mosseri, spoke to the new products the social platform could create to get beyond their many problems:

> The truth is we've learned things since the election, and we take our responsibility to protect the community of people who use Facebook seriously. As a result, we've launched a company-wide effort to improve the integrity of information on our service. It's already translated into new products, new protections, and the commitment of thousands of new people to enforce our policies and standards… We know there is a lot more work to do, but I've never seen this company more engaged on a single challenge since I joined almost 10 years ago.[26]

One of Facebook's solutions was what they called "trustworthiness ratings." To crack down on fake news and misinformation, Facebook would give each of its users a trustworthiness score. This score would be assigned to each of us based on our previous behavior, such as whether we flag posts as fake when the algorithm says they're not.

So what could possibly go wrong with a trustworthiness slug-out between humans and algorithms?

If this trustworthiness score reminds you of what the Chinese have instituted to suppress dissent, that's because it is similar. Or exact, even. Beginning in 2018, every Chinese citizen was assigned a score which is used to determine what they can and can't do as citizens. Facebook doesn't have that much power over our lives (yet), but they do have power over what a lot of people say and think. Here in America, so far, this rating system is apt to be short-lived, and, we hope, shamed into the dustbin for what it is: authoritarian bunk.

Meanwhile, the social platform has been spending hundreds of millions to convince people that all is swell with Facebook. A big slick advertising campaign, featuring a video called "Facing Facts," may be one of the finest productions of the year—may even get an Oscar nod in the Shorts category.

But all of the gauzy overlays do little to truly address the rampant fraud, partisanship, hate speech, identity scams, and fake news spilling across the platform. All of this is clearly growing in people's minds as a deeply troubling problem. In surveys taken following Zuckerberg's testimony, the Edelman Trust Barometer found only 30 percent of Americans trusting social media as a source for general news and information, down from 35 percent in 2015.[27]

And so it's no wonder...

DOJ, FBI, SEC, FTC—All Now Investigating

Facebook may also have set a record: the first company to be investigated by four government agencies at once. The Federal Trade Commission has been joined by the Department of Justice, Federal Bureau of Investigation, and Securities and Exchange Commission, all looking into Facebook's "actions and statements" since the first privacy debacle in 2007. This rare multi-agency inquiry is looking to determine what Facebook has been hiding from both users and investors.[28]

Within hours of Zuckerberg completing his latest testimony on Capitol Hill, there were two new bills proposed in the Senate. Each of them was intended to exact new penalties for data breaches, and to require companies like Facebook to make it easier for its users to opt out of ad tracking. It's a start, but only that.

Around the world, regulators are beginning to realize that our personal data is too important to be left to companies that only profit when they manipulate it against us.

In June of 2018, California's governor signed into law the Consumer Privacy Act to give people more control and "the right to tell a business not to share or sell your personal information."[29] A month earlier in Europe, a new "General Data Protection Regulation" (GDPR) began requiring that advertisers ask upfront for permission to capture or use your personal data.

If these piecemeal reforms can make a genuine difference, we're all for them. But Zuckerberg and company have an eleven-year history of blowing smoke at legislators, then going back to doing whatever they please. Their business models and bonus checks are perversely tied to manipulative systems, so they have every reason to find new ways around the legislators. Therefore, the job of government should be to get the right incentives in place that protect the consumer.

History shows that laws work best when aligned with proper incentives.

The Leftist Dog Whistle of "Hate Speech"

Thanks to modern technology, much of it originating in Silicon Valley, it's now possible to be a fly on the wall overhearing almost any conversation. And so…

"I got it: Anything we disagree with we'll just label as 'hate speech'—that'll keep them backpedaling."

—OVERHEARD AT LIBERAL LEADERSHIP
GATHERINGS SINCE 2014

"It's not their fault—they've been arbitrating cultural norms for so long, they truly believe they are 'love.'"

—OVERHEARD AT CONSERVATIVE LEADERSHIP
GATHERINGS SINCE 2014

Or so we imagine. But moving from the sublime to the sadly ridiculous, we are here publishing for the first time…

Field Guide to Leftist Dog Whistling

Liberals in general = **love**	Conservatives in general = **hate**
Hating the president = **love**	Loving the president = **hate**
Expressing pride in your cultural history if you're LGBT, female, a refugee, or black = **love**	Expressing pride in your cultural history if you're a Christian, male, heterosexual, or white = **hate**
Establishing sanctuary cities for illegal immigrants = **love**	Defending national borders from illegal immigration = **hate**
Ticking off a long list of America's failings = **love**	Proud to be an American = **hate**
Want tougher gun laws = **love**	Stand with the Second Amendment = **hate**

We've taken a crack at identifying leftist dog whistles in the wild because, frankly, we've never seen it put this way. We often see leftists carrying on about right-wing dog whistles of which there are, they insist, an abundance. But not so much the opposite. So, in fairness, we are supplying this handy guide, not only as an exercise in political correctness in the literal sense, but also because this thinking—these dog whistles, if you will—are now being hard-coded into the social platforms.

They are being used by Facebook, Google, and Twitter now to justify banning political speech that they happen to disagree with. Leftists who run the tech giants today almost universally believe that any opposition to their agendas must be rooted in "hate," and therefore legitimately silenced.

Their algorithms are thus being weaponized to furtively silence online speech they disagree with.

Yes, weaponized. For with the power to define "hate" and use it as justification for censorship comes the power to dictate which opinions get shared online.

And yes, these "hate" labels are being quite arbitrarily assigned, failing any legitimate test of logic. They are being used to shadow ban, blacklist, or outright silence a spectrum of the free ideas and opinions of a nation.

Hate speech simply becomes whatever happens to contradict the latest narrative of political leftists. For instance, the leftists were vehemently anti-illegal-immigration in the 1970s when it suited them, as we'll discuss elsewhere. But it no longer suits their electoral ambitions, so now they are vehemently pro-illegal-immigration. So a number of the voices falling on that part of the spectrum of color are being filtered one way or another.

Leftists no longer believe dialogue or debate can serve any useful function in achieving their goals. And granted, the nation is divided on loads of issues and trying to change opinions is like pulling dinosaur bones out of tar pits. But that's also one of the great things about an open society. We can pull and pull; we can have that debate.

But having given up on dialogue and debate, the leftists have turned to the tools of censorship, a *Fahrenheit 451*-like treatment of online words and pictures.

Or as the fly on the wall might have heard it:

Why fret over any finer distinction between the Heritage Foundation's call for tighter monetary policy and the National Policy Institute's call for an all-white nation. Just lump all those right-wingers into the category of 'nasty alt-right haters' that needs to be censored. Period. End of debate. What's to be discussed? We all agree on this stuff, right? Show of hands? Ah, heck, we know how we feel. Passes with unanimous consent. Now pass me a quinoa doughnut...

Facebook got in line behind the censorship wagon in January 2018 by announcing a "philosophy reboot." Henceforth, the social platform would be boosting "higher-quality news publishers" in its rankings, using surveys to identify "trusted sources" and prioritizing their content ahead of clickbait, propaganda, and fake news.[30]

Facebook predicted that the average user's feed would go from comprising about 5 percent news stories to about 4 percent—or very little shift, at all.

It was tough to know what impact sites would see from this shift. Other than saying that the stories people saw in their News Feeds would shift a little, Facebook was keeping secret what algorithm tweaks it had made to accomplish its goals, and what the impact would be.

We do understand why Facebook was being secret about the mechanics of the change. It didn't want to give away information that could help companies game its algorithms. But that secrecy left the company vulnerable to criticisms of bias. Extremely vulnerable when one thought through how those algorithms were being tweaked by engineers who quite publicly boast of hating the "haters"—that is, those they disagree with.

Maybe some of the engineers could show up for work and separate their personal feelings from their professional workflow. But the evidence suggests otherwise.

Exhibit A: Damned by Your Associations

When politicians run into trouble, and things look bad, they know there is an activist group they can run to. Both sides have one, but on the left the go-to group is named Media Matters. It's led by Angelo Carusone, and it's basically an attack dog specializing in *ad hominem* attacks on political opponents, as needed.

And so, with the leftist Facebook facing a new round of accusations, they sent word to unleash Carusone in full attack mode.

Within days, Carusone was out in the press insisting that the Facebook algorithm change had been applied evenly across its universe of publishers, that any suggestion otherwise was "a faux scandal cooked up by conservatives" over unsubstantiated allegations of bias.[31]

So that was the first thrust. Then Carusone went on to call the substantiation "demonstrably false" since conservative sites such as *Daily Wire* and *Western Journal* were outperforming liberal websites. Ipso facto, the conservative claims of bias were "demonstrably false." There could be no other reason conservative sites were outperforming liberal sites, right? Carusone's twisted logic was good enough for his liberal audience, no doubt. For all they wanted was to feel righteous in their predispositions.

Exhibit B: Laying Out the Censorship Plan at a Private Retreat

Back in January 2017, a private retreat was held at the posh Turnberry Isle Resort in Aventura, Florida—as uncovered by the *Washington Free Beacon*.[32] More than a hundred leftist operatives and money handlers were on hand to, as Media Matters founder David Brock said, "kick Donald Trump's ass."

Much of the group's funding was coming from George Soros. Top activist groups included Media Matters along with American Bridge, Shareblue, and Citizens for Responsibility and Ethics in Washington. The objective: Map out strategies to take down the president.

Strategies included:

- Laying the groundwork for impeachment.
- Combating any and all misinformation from the Trump Administration.
- Gaining Democratic control of Congress in the midterm elections.
- Overwhelming the Trump Administration with frivolous lawsuits.
- Using digital attackers to make up stories about Trump and Republicans.
- Partnering with Facebook to combat "fake news."

Brock claimed at that retreat to have "access to raw data from Facebook, Twitter, and other social media sites," allowing them to "systemically monitor and analyze this unfiltered data."[33] If this claim was accurate, it suggested that Media Matters had a special relationship with the social platforms that other publishers did not.

Whether true, or whether Brock was just bloviating in front of potential donors, Media Matters operatives did say that they had met with Facebook to talk about how best to crack down on fake news. Media Matters handed over "a detailed map of the constellation of right-wing Facebook pages that had been the biggest purveyors of fake news."[34]

Media Matters gave Google the same "blacklist."

Following these meetings, coincidentally, Facebook and Google both changed their newsfeed algorithms, saying they were only trying to fight "fake news" but having the result of tanking traffic to many conservative sites.

President Trump himself was affected; his engagement on Facebook dropped 45 percent.

Gateway Pundit studied the effects and found that Facebook had eliminated 93 percent of the traffic of top conservative news organizations.[35]

Western Journal also studied the data and found the average liberal-leaning publisher saw a 2 percent increase in web traffic following Facebook's algorithm changes, while the average conservative-leaning publisher saw a 14 percent loss of traffic.

Exhibit C: Facebook's Own Employees Speaking Honestly

Any attempts to camouflage the truth of Facebook bias fell apart when several former Facebook employees—people who had worked as news curators—went on record saying that, yes, they had routinely manipulated the trending news feature to exclude topics of interest to conservatives. In one curator's words: "It was absolutely bias.... It just depends on who the curator is and what time of day it is."[36]

JD Heyes, editor-in-chief of the *National Sentinel,* sums it up best: "Mark Zuckerberg should just come out and say it: 'If you're pro-America, pro-President Donald Trump, conservative—or any combination of those things—you're not welcome at Facebook.' Because in essence, that's what they're already saying."[37]

Net net: When two social platforms are together filtering up to 90 percent of the news people are seeing online—that's an effective monopoly, or duopoly. And if they are truly guilty of censoring, that's not okay.

Turning Algorithmic Artillery on the Homeland

Tech Giants Google and Facebook are:
- **Purging conservative content**
- **Hiding conservative stories**
- **Shadow banning conservative news**

First Casualty: The Tea Party News Network

We take no pride in being the first casualty in the Social Wars, as future historians might call them. One of the authors, Todd Cefaratti, had built the nation's largest Tea Party organization, TheTeaParty.net. In just five years, this organization of independent thinkers grew into a powerhouse of eight million dedicated activists. From there we expanded into a media group, the Tea Party News Network. Within a month of our November 2012 launch, TPNN was receiving more than sixty million page views per month.

Facebook played an instrumental role in these achievements. The social platform gave us a cost-effective channel for reaching millions of viewers with a message that resonated. We would write stories and place them in the Facebook News Feed. This is what's known as "user-generated content," and it's the core genius behind Facebook's hypergrowth.

Our stories prompted millions to consistently go to our TPNN website and read our content that was not appearing in the "mainstream" media. We were over-the-moon delighted with our Facebook partnership.

So we went ahead and began purchasing ads on Facebook to increase our followers—those ads that run down the right column of your News Feed. These ads, Facebook said, could boost the reach of our stories to many more people. We invested many thousands of dollars in this Facebook marketing program. It allowed us to target people who were most likely to appreciate our message, and to encourage them to follow our pages. It worked.

It worked well beyond our expectations for the first five years.

People were reading our news stories in record numbers. And they were engaging with these stories—commenting on them and sharing them with their friends. We could see the traffic of people visiting our website in real time using the Google Analytics tool. At any given time, the tool told us, 10,000 to 20,000 unique visitors were reading stories on our site. Huge numbers for a political site.

Overnight, everything changed.

On a Sunday morning in February 2015, we woke up early and, as usual, looked at our Google Analytics tool. Our traffic had fallen off a cliff—from a steady and dependable average of 10,000 to 20,000 visitors—all the way to down to 800. It was one of those days you just wanted to go back to bed.

How could we lose maybe 19,000 visitors overnight? We were posting the same types of stories we always had. Our audience was also a strong and loyal one—we knew from all the cards and emails received. What had happened?

Our first thought was, we'd been hacked! Desperate to figure it out, we hired some expensive IT consultants to review each of our websites. They went over everything with a fine-tooth comb and came back to us with what we first considered good news: They found no hacks.

So we contacted our account manager at Facebook to see if we had done something wrong that caused them to throttle our traffic. We had several calls with our Facebook rep, and the company's tech team as well, to get to the bottom of things. Every call resulted in a version of the same answer: *"Facebook hasn't done anything to your traffic."*

They also referred us to their "best practices" guidelines. We pored over them word for word. We had done nothing wrong, nor even out of the ordinary and never received so much as a warning from Facebook. Indeed, we had been posting the same flavor of content since the beginning.

After two days of frustration, our traffic recovered—miraculously, it seemed—to the normal 10,000 to 20,000 numbers. Oh, we were relieved! It must have been, we decided, some algorithmic snafu at Facebook. End of nightmare.

One day later our traffic crashed again.

We were back to a low of 800 visitors. And there we stayed…flat as a pancake. Still, we didn't believe it could be anything deliberate. We checked to see if a new algorithm release from Facebook had triggered the cliff dive (when a social platform changes its underlying algorithms, all outside parties are impacted somehow). We found no notice of any such release on all the technical sites that track major code changes.

We were stuck with lots of questions, but no answers.

At this point, we checked in with our friends at similar websites. Had they experienced any similar problems? The answer was no. In time, that answer would change; but right then, we appeared to be the first casualty in a war zone we didn't know existed.

After spending five years building an impressive conservative audience, we were cut off from them, as if the tide had gone out and stayed out—and we were stuck on an island no longer close to shore. No voice, no outreach, no connections. All we did have, by process of elimination, was the hard fact that only one entity (Facebook) had the power to cut us off from the thousands of people who wanted our message.

It was then, stuck on an island when we thought we were living in a free country, that the future of social media became clear.

We would not be alone for long. We suspect that Facebook started with the Tea Party News Network because of our success and the Tea Party brand. Our name and our reach made us an easy initial target. They were starting a war with free speech and conservative America.

And as in all war, it was going to get worse.

Facebook's Guns Take Aim at the President

The audacity of anyone trying to kneecap any President's speech is thoroughly astounding. One can just imagine the screeching charges of racism by leftists if the shoes were switched and conservatives controlled 90 percent of the news people see online, and Obama was snuffed out! All hell would break loose—as it should. But in the situation as it is, it's pretty much crickets on the left. Or weak denunciations at best, with few honest actors.

President Trump gets that. Now that it's no longer a mystery, he knows he's at war. So the president is looking at the best action to take against social platforms that remove content from conservatives. His tweets @realDonaldTrump:

"Social Media is totally discriminating against Republican/Conservative voices"[38]

"Censorship is a very dangerous thing & absolutely impossible to police. If you are weeding out Fake News, there is nothing so Fake as CNN & MSNBC, & yet I do not ask that their sick behavior be removed. I get used to it and watch with a grain of salt, or don't watch at all."[39]

"Social Media is totally discriminating against Republican/Conservative voices. Speaking loudly and clearly for the Trump Administration, we won't let that happen. They are closing down the opinions of many people on the RIGHT, while at the same time doing nothing to others."[40]

"Too many voices are being destroyed, some good & some bad, and that cannot be allowed to happen. Who is making the choices, because I can already tell you that too many mistakes are being made. Let everybody participate, good & bad, and we will all just have to figure it out!"[41]

Also in Their Sights: The "Roe v. Wade" Movie

Dr. Alveda King is the niece of Dr. Martin Luther King, and while some lesser women might have ridden the great man's fame and settled back, she took the harder path of rigor and education. She earned her master's degree in business management from Central Michigan University, received an honorary doctorate from Saint Anselm College, and became the director of civil rights for the unborn at Priests for Life. Her credentials are noteworthy. But not good enough for Facebook, apparently.

King's latest project is producing the pro-life movie *Roe v. Wade* about the Supreme Court's 1973 decision, featuring an all-star cast with Jon Voight, Stacey Dash, Corbin Bernsen, Steve Guttenberg, and other accomplished actors. It's due for release in 2019. King and her team turned to Facebook to promote the film, and to help crowdfund it. But Facebook decided to "pull down" the ads she had already paid for.[42]

Why would Facebook yank paid ads when they're in the paid-ad business? King says Facebook does not "want the message of the injustice of abortion broadcast" even though that decision is a "violation of religious freedom... discriminatory."

Facebook would not comment on their behavior. But King believes Facebook is choosing to censor the story because it tells how the events leading up to the Supreme Court decision were grossly manipulated and misrepresented—a story few know. It's story that could infuriate abortion advocates who don't want any challenge to the narrative they've burned into the national imagination.

Nick Loeb, an actor and the film's co-executive producer, says he may file a lawsuit against Facebook for breach of contract:

> We did a lot of our promotions through Facebook, as most people do to do crowdfunding. Facebook sort of controls all of the ability to raise awareness, and for crowdfunding. And they blocked us from sharing our posts with our friends. We even then went and paid for ads, and they blocked us even after we gave them money. They blocked us from sharing the ads.[43]

Loeb may have a strong case. He paid for the Facebook advertising campaign but was not allowed to receive the full benefit of the campaign as detailed in Facebook's terms and conditions.

Just as her uncle fought to tell the truth about racism in America, Dr. Alveda King is fighting to tell the truth about abortion in America. Both are worthy struggles. Both are important. People can disagree about these issues and the facts surrounding them. They can argue it to the bone—because abortion is a sensitive subject. But people cannot be allowed to censor the discussion. Not in America.

Yet they are being allowed to.

Next: Targeting Books That Dare Criticize Obama

Matt Margolis's first co-authored book was *The Worst President in History: The Legacy of Barack Obama*.[44] It was well-reviewed, even acclaimed by some. Like King, Margolis is contributing to the national conversation, and poses no threat to the republic. Yet when he released his latest book, *The Scandalous Presidency of Barack Obama*, he was temporarily banned from promoting it on Facebook.[45] He was trying to promote his book on the social platform, as most authors do—because when allowed to, it can be very effective.

But Facebook arbitrarily, and without explanation, silenced Margolis by banning him from the platform for six days. Why six days? Facebook has offered no explanation. But the social giant may know two things:

- Their actions skirt legality, so they have to have an "Oops, sorry, didn't mean that!" defense always at the ready.
- The prime promotional time for a book is the first week of its release—so denying that promotional period to an author can be a crippling blow.

Margolis explains what he thinks happened to him (as told to *PJ Media*):[46]

> I hate to think I'm being censored, and hope that's not the case... but Facebook's reputation makes that possibility hard to ignore. We're seeing companies like Facebook, Twitter, and Google actively censoring conservative ideas on their platforms. It is hardly far-fetched that big-tech censorship wouldn't include a book that doesn't fit the narrative they support.

When you think about the vast complexity of a platform like Facebook, with billions of people posting and tapping and clicking around, it's hard to imagine the company's teams and computers could even keep track of all the content they might wish to censor. There's just so much going on, and so much that's automated.

How does censorship even happen?

In the Social Wars, the left and right wings are locked in constant battle with each other. In this particular case, radical leftist trolls spent time searching for conservative messages and dreaming up inventive ways to discredit them (branding them hateful, racist, bigoted, patriarchal, discriminatory, etc). It's done in the same narrow partisanship that some of their counterparts on the radical right employ. It's warfare. But it's warfare the platforms like Facebook have enabled, so they must now, somehow, make free speech work in it.

Advancing solely the Big Tech Tyrants' political views is not only an unacceptable solution—it's downright dangerous to America's democracy.

In Ray Bradbury's *Fahrenheit 451*, a liberal world found their *cri de coeur*—we can never allow an autocratic government or technology platform to ban books lest we forfeit our essential humanity. But isn't banning the promotion of a book the same as banning the book itself? Strip away all the politics, and the answer is yes.

Freedom of speech should never be freer than in the pages of a book. A book is an author's perspective, backed by facts and data, or opinion and conjecture, or lies and innuendo, or a combination of all three. That's the author's choice.

It's up to the reading world and its book-buying choices to judge the author's work through sales and critique, unthrottled by government or industry.

It's not up to the platforms to interpose themselves between authors and audiences. Yet they are.

Cutting Deepest With News Stories That Go "Poof"

For some time now, Facebook's "Trending News" section has been the most powerful real estate on the internet—because at any given moment this is what Facebook's 167 million US users are reading. News stories popping up on Trending News on users' apps or desktop have been pushed there by Facebook's curators, sort of.

Technically, according to Facebook, these curators had a cushy job for more than a decade. Because Facebook used an "impartial algorithm" to determine which news stories were gaining in popularity—that is, trending—so the stories would be pushed out to users in a certain priority. The algorithm would only push stories that the user would presumably be interested in seeing—based on the stories the user had "liked" in the past. It was a mostly automated system, one Facebook heralded as the future.

Curators would merely write snappy headlines and summaries for each news story and include links to news sites. A final duty for curators—they would ensure that topics were, in fact, trending in the real world and not junk topics, duplicate topics, hoaxes, or of questionable provenance.

It all sounded reasonable on its face.

Until a Facebook employee who had worked at the social platform as a curator in 2014 and 2015 stepped up and admitted:

> I'd come on shift and I'd discover that CPAC [Conservative Political Action Conference] or Mitt Romney or Glenn Beck or popular conservative topics wouldn't be trending because either the curator didn't recognize the news topic, or it was like they had a bias against Ted Cruz... Depending on who was on shift, things would be blacklisted or trending. I believe it had a chilling effect on conservative news.[47]

This former curator asked to remain anonymous, citing fear of retribution from Facebook. He was so troubled by the omissions that he kept a running log of them at the time and turned that log over to *Gizmodo*, which broke this story and vouches for its authenticity. *Gizmodo* is not a "political" site. It focuses on technical design and is read by Silicon Valley insiders. There was no attempt at scoring political points here; just reporting a story of interest in the Valley, and beyond, as we would soon learn.

Included in Facebook's suppressions were stories involving:

- Former IRS official Lois Lerner (who had harassed conservative groups)
- Wisconsin Gov. Scott Walker
- Popular conservative news aggregator the *Drudge Report*
- Chris Kyle (the former Navy SEAL who was murdered in 2013)
- Former Fox News contributor Steven Crowder

These were hot topics at the time, and they were deep-sixed—actually omitted from the Trending News pages of Facebook users. A damning indictment—if the former curator was telling the truth. Soon there would be corroborating evidence. A second curator stepped forward to say that yes, the social platform had an aversion to right-wing news sources:

> It was absolutely bias. We were doing it subjectively. It just depends on who the curator is and what time of day it is…Every once in a while a red state or conservative news source would have a story. But we would have to go and find the same story from a more neutral outlet that wasn't as biased.

Stories covered by conservative outlets—such as *Breitbart, Washington Examiner, Western Journal, Drudge,* and *Newsmax*—that were popular and trending according to Facebook's algorithm were nonetheless suppressed unless leftist sites like the *New York Times*, the *BBC*, and *CNN* covered the same stories.

Gizmodo sought to tell an honest story and did seek out other former curators. Some said they did not willfully bottle up conservative news, so it's fair to conclude that not every curator was exercising a blatant leftist bias. But with these ethical lapses exposed on *Gizmodo*, Facebook found itself in damage control mode once again. Facebook's vice president of search, Tom Stocky, whipped out a post: "We take these reports extremely seriously and have found no evidence that the anonymous allegations are true" and in fact would not be "technically feasible."[48]

A Facebook press release followed quickly on Stocky's heels:

> Facebook is a platform for people and perspectives from across the political spectrum. Trending Topics shows you the popular topics and hashtags that are being talked about on Facebook. There are rigorous guidelines in place for the review team to ensure consistency and neutrality. These guidelines do not permit the suppression of political perspectives. Nor do they permit the prioritization of one viewpoint over another or one news outlet over another. These guidelines do not prohibit any news outlet from appearing in Trending Topics.

A few hours passed, and Facebook issued a somewhat more nuanced statement:

> Our review guidelines for Trending Topics are under constant review, and we will continue to look for improvements. We will also keep looking into any questions about Trending Topics to ensure that people are matched with the stories that are predicted to be the most interesting to them, and to be sure that our methods are as neutral and effective as possible.

Ah, the old, "We didn't do it, or if we did, it wasn't us" defense!

It should come as no surprise that (a) Facebook employees have biases that inform the editorial decisions they make, and that (b) Facebook would deny those biases affect the makeup of their product. But when Facebook would like to be thought of as an impartial mirror of the *vox populi* (voice of the people) and has the power to influence what billions of people see and has even openly discussed whether it should use that power to tilt a presidential election…we have a problem.

Is Bias Confirmed by Inability to Take a Joke?

Humor can be a wonderful truth serum to unmask people's motives and reveal their primary colors and biases. Consider this story that ran in the *Babylon Bee*.[49]

Facebook's curators could be excused for not knowing the *Babylon Bee* is a Christian satire site, founded by *Adam Ford*. But when those curators found a story on Facebook headlined "CNN Purchases Industrial-Sized Washing Machine to Spin News Before Publication," you would expect them to recognize satire when they stumble over it.

Apparently not.

Mr. Ford's post was tagged with a stern warning from Facebook. Dogged in their efforts to protect users from "fake news," the curators flagged the story and submitted it to their so-called "independent fact checkers" at *Snopes* who found the story to be "FALSE."

> Claim: CNN invested in an industrial-sized washing machine to help their journalists and news anchors spin the news before publication.

> Snopes Rating: FALSE!

Armed with independent verification that CNN had never, not even once, run their news through a washing machine, the intrepid curators at Facebook promptly notified Mr. Ford of his transgression.

Upon receiving this warning, Mr. Ford could have laughed it off. He is, after all, a humorist. But he was also told that repeat offenders "will see their distribution reduced and their ability to monetize and advertise removed."

Amazing! Since common sense had clearly left the building, Facebook felt the need to further threaten the *Babylon Bee* with demonetization (losing revenue-generating advertisers) as a "repeat offender" charged with peddling fake news.

Eventually, egg-faced, Facebook's Lauren Svensson did reach out to the *Babylon Bee* with something of an apology, acknowledging "there's a difference between false news and satire."

Was this a sincere apology or a cover-up? Corporate CYA'ing or insight into the social beast? Do Facebook's curators and fact checkers hold conservative Christian sites in such contempt, or alternatively hold *CNN* in such reverence, that they put the *Babylon Bee* through a wringer of their own? Decided to "rinse and spin" the Christian publisher…because they could?

Unfortunately, we would see again that only rarely is Facebook's censorship so silly.

Using Fake News to "Fight" Fake News

While Facebook censors conservatives under the guise of "weeding out" fake news, they have no such concerns about a Silicon Valley couple that used their platform to raise nearly $20 million—and used fake photography to do it.[50]

It was a big fundraiser being run on Facebook's pages, with high-profile sponsors and lots of people seeing it—not something that might have snuck under the radar. The fundraising was being done for the charity

RAICES—the Refugee and Immigrant Center for Education and Legal Services. The couple behind it posted a story featuring a photo of a toddler crying at the feet of immigration officers, painting a picture of the Trump administration tearing young girls from their parents at the border. But the photo was fake. The child in question had been with her mother at the time; the two were never separated.

Time magazine also used the same fake photo, taking a giant irresponsible step further by Photoshopping the young girl together with the president in a sea of red.[51]

Time's page was not censored either, despite being obviously fake. It should be noted that John Moore, the Getty Images photographer who took the picture, was quick to notify *Time* that their caption was incorrect, and *Time* was forced to issue a correction about the nature of the little girl's distress:

> "The original version of this story misstated what happened to the girl in the photo after she [was] taken from the scene. The girl was not carried away screaming by U.S. Border Patrol agents; her mother picked her up and the two were taken away together."

So at least *Time*, though not Facebook, made some effort at professional journalism—admitting to the "hit job" after the fact when it didn't matter, the damage was done.

Conservative News Sites Actively Harassed

Another popular Facebook page is the "Health and Global News" page run by *Allen Muench*. With more than 13,000 "likes," this page appeals to independents who don't typically vote. For his efforts at building a Facebook following and creating and posting his content, *Muench* was censored or suspended nearly twenty times in 2017.

Facebook also sent him error messages, as reported in the *Daily Signal*.[52] So what kind of erroneous stories was Muench publishing at "Health and Global News" that would merit suspension—especially given the Wild West, anything-goes publishing culture of Facebook? How inflammatory, incendiary, obscene, or hurtful were these stories? Here are two culprits:

- Article titled "Two Afghans Arrested for Raping 16-Year-Old on German City Street"
- Video of the American flag

That's right, a normal, everyday story about the terribly unfortunate side of unrestricted immigration, and the American flag. The kinds of stories and videos that could be found on any number of thousands of Facebook pages which were *not* suspended. So why did Muench run into such trouble?

He believes his troubles stem from Facebook's selective enforcement of their Community Standards policies: "Facebook seems to enforce their policy more on conservatives than liberals."

Folks over at the conservative Catholic News Agency can relate. In the summer of 2017, they discovered that Facebook had blocked more than twenty of their pages over a twenty-four-hour period.[53] Some of the blocked pages had millions of followers and were in demand by Christians. Yet they were banned.

What would possess Facebook to ban a page named "Jesus and Mary" with its 1.7 million followers? Facebook accused the page of "suspicious activities." Keep in mind, the page's main cover photo was of the sacred hearts of Jesus and Mary. It stretches the imagination as to what kind of "suspicious activities" Facebook suspected.

At first, the Facebook banning was said to be temporary, so life could go on for this site—which depends on traffic to pay its bills. But when the site administrator returned a week later, he was told that his page with its 1.7 million followers had been "disabled."

All his work, ruined.

Reporting on this rank censorship, the Catholic News Agency talked with *Kenneth Alimba*, who runs the six-million-strong "Catholic and Proud" page. He says his page, and two dozen others like it, are being deliberately censored by Facebook: "They've fought and continue to fight anything Catholic and conservative."[54]

Even the pages of *well-known* Catholic leaders have been unfairly targeted. "Fr. Rocky" is run by Father Francis J. Hoffman, executive director of Relevant Radio. This Facebook page boasts 3.5 million "likes." He, too, has been banned.

Why ban decent religious sites and then ignore the broad swath of nasty, distasteful, and violent bad actors that litter the Facebook universe?

As a technology platform that has become the world's gatekeeper of information, Facebook has a responsibility to be "hands-off" unbiased, but is not required to under current law. So if Facebook is abusing its responsibility, and cannot keep its hands off other people's information, then the current law is no longer working and should be addressed.

One would think that Facebook would ban as few sites as possible—say, only banning the sites breaking some law. Because if a site is receiving traffic, that's money in Facebook's pocket. So once again it appears that the only logical explanation for Facebook's behavior is that an agenda has formed. Facebook's execs have decided, either on their own, or as a result of the constant prodding from leftist trolls, that conservative or Christian points of view must be silenced.

The political becoming the personal, superseding the financial, rationalizing censorship. Or as the folks at Facebook might think of it…

> We do control the flow of modern information. We're really untouchable, you know. We have no reason to fear reprisals of any kind. Not from customers, competitors, government regulators, anyone. So lifting up our own prejudices to the level of policy is a small price to pay—a rounding error, really, in the corporate P&L, barely denting ROI. The real ROI is the immense joy we feel here at Facebook for being able to so effectively silence the voices we loathe so deeply.

Going back to the beginning, the original appeal of the social platforms was that you didn't have to rely on the judgment of professional editors. A good thing, presumably, because all humans come equipped with bibs and bobs of bias. So Facebook was pioneering a brave new way forward, but not an easy way. Fraught, in fact—as we learned. So later in 2018, Facebook fell back on their plan B and, whether they intended it or not, they basically doubled down on the problem. Plan B involved the hiring of 10,000 additional curators to disinfect the conversation on Facebook pages.

"The 10,000," as they may someday be called, won't have it easy. They begin on the retreat, as a lot of people have a beef with Facebook now. Do they secretly know that they have no chance ultimately of succeeding? For when you have 2.2 billion people on a single platform, each of these 10,000 is responsible for 220,000 Facebook pages. Outnumbered 220,000 to one. Even with all the tools Facebook can put in their hands, how can 10,000 police 2.2 billion? Good luck with that.

It's ultimately an issue of scale. Some things only become problems when they grow too large. Think obesity, for example. Having grown large, they can only be corrected by being made small again.

When Shadow Banning Becomes Criminal Behavior

The social platforms are private companies, free to run their operations as they choose. Banning, shadow banning, and blacklisting conservatives is generally within their rights under current law. What they're doing is unfair, but a lot of things in business are unfair. However, there are legal ramifications to certain kinds of discrimination.

The first ramification is straightforward:

When an organization pays a fee to a social platform in order to increase their reach and exposure on the platform, a contract has been entered into. And if during that contract the platform purposefully clamps down on the organization's reach or exposure, then the platform is in breach. That is what has happened.

Facebook and Google have both taken money for their services and not delivered.

It doesn't matter if the platforms came back later and apologized, insisting that they crimped a client's traffic "by accident" or by some "algorithmic snafu." If the platforms have taken money, and not delivered—it is fraud. Some would even call it bait-and-switch.

The courts have traditionally called it criminal. (And right now, several such cases are in court, as we will see...)

The second ramification is long and twisted:

It begins with the 1996 Communications Decency Act. This act basically says communication platforms are not legally responsible for what other people publish on their sites. And today's social media platforms are today's communication platforms, just like telephone's dominated in the 20th century. This distinguishes them from publishers, like newspapers, which are held responsible for their decisions. If a newspaper publishes a libelous story, it can be sued. If it infringes on a copyright, it can be held liable for damages. So the burden of being a publisher is in deciding what to publish and what not to. The social platforms don't have that burden.

The Communications Decency Act was declared unconstitutional by the Supreme Court, but it had an amendment known as "safe harbor" which survived the court's decision. Known as Section 230, this safe harbor plays out in two parts:

1. **Internet platforms such as Facebook do not need to police what its users say or do and cannot be held liable for the speech of its users.**

2. **If a social platform chooses to police what its users say and do, it does not lose its safe harbor protection in the process.**

So choosing to purge or shadow ban some content does not turn a platform into a publisher with liability concerns. As Harvard law professor Rebecca Tushnet put it, they get the power without the responsibility: "Current law often allows internet intermediaries to have their free speech and everyone else's too."[55]

Again, not fair, but technically legal. However, how does that change when you add up the reach enjoyed by Facebook, Google and Twitter, and see that they control up to 90 percent of the information that people see online? Here is where the dynamic of their Section 230 protection changes.

Their dominant hold on information has clearly stretched the spirit and intent of the 1996 law, and brought into play another court decision fifty years before that. Back in 1946, in Marsh v. Alabama, the federal court decided that in a company town, where a private company owned the sidewalks and streets, they could not stop a citizen from handing out religious literature on those sidewalks and streets. People could not be deprived of the liberties they were guaranteed by the First and Fourteenth Amendments to the Constitution.[56]

Today the social platforms are the sidewalks and streets of information flow, the functional equivalent of those old company towns, so the courts have precedent to forbid the platforms from discriminating.

And there is no shortage of evidence to support such court cases, including...

- Elizabeth Heng, who ran as a Republican in California's 16th Congressional District. Facebook blocked her from posting a video which included her family story of having survived the Cambodian genocide. In contrast, another minority woman, Alexandria Ocasio-Cortez, came out of nowhere to win the Democratic primary and seat in New York's 14th Congressional District. Ocasio-Cortez not only ran all the videos she liked on Facebook but appeared to be getting viewership "boosts" from Facebook. Heng lost her race...and Ocasio-Cortez is now the new congresswoman from New York.
- Alex Jones runs the website *InfoWars*, which has gained a lot of media attention in recent years because Jones has been known to make some very controversial claims. In a true free speech society like America,

people should be allowed to communicate their opinions, no matter how controversial they may seem to some and it is up to the receiver of that information how it is interpreted. However, people should still have the right to receive any information out there if they choose to. That is the definition of free speech. For years Jones has published his content including the controversial claims on the big social platforms and enjoyed a massive following. Then for some unknown reason after five years of the exact same type of content, Facebook, YouTube, and Google decided he no longer met their "Community Standards" which he had met for years and banned him and *InfoWars* from their communities.

- Jones's circumstance has another dimension that's relevant. His *InfoWars* is itself a media operation, a competitor to the social platforms. When he was banned from all the platforms including Twitter, Facebook, Google, Apple, and You Tube, the joint bannings could be construed as an attempt by several platforms to knock off a competitor. The legal term is "conspiracy in restraint of trade," and it is illegal.

- *Vice* is a racy publisher for young men. Its co-founder, Gavin McInnes, along with his previous Proud Boys organization who called themselves "Western chauvinists," have been banned from Twitter. Meanwhile, any number of "Western disparagers" run free on the platform. So Twitter's decision-making is either quite arbitrary, or it just doesn't like people who like America.

These are a few of the many special cases of discrimination we'll visit in these pages.

Using "Fact Checkers" to Appear Impartial and Unbiased

When the social platforms came up with plan B—to use impartial fact checkers to combat all the misinformation on their sites, they must have felt that they dodged a bullet. Here at last was a way to muzzle their detractors with the irrefutable logic of third-party, independent verification of "the truth."

So in 2018, Facebook hired 10,000 fact checkers. They relied on a variety of sources for these humans who would have the unenviable task of doing what machines used to do—wading through billions of daily posts to "find truth and quality." To put their task into perspective, in the first quarter of 2018 alone, Facebook uncovered per their report:

- 837,000,000 spam, false advertising, or malicious links
- 583,000,000 fraudulent account actions
- 21,000,000 obscenity violations
- 3,400,000 graphic violence
- 2,500,000 hate speech
- 1,900,000 terrorist propaganda

Into this sewage the fact checkers were tossed. They came from many "news" organizations including *ABC News,* the Associated Press, *FactCheck. org, PolitiFact, Snopes,* and the *Weekly Standard* (closed in December 2018). All but one of these, the *Weekly Standard* (which just so happened to be a strong critic of President Trump), are known to have a strong liberal bias. So it's no wonder Facebook believed these would be the ideal arbiters of truth and quality in the Facebook universe.

Facebook also turned to the Southern Poverty Law Center (SPLC) for advice to "inform our hate speech policies."[57] This sounded like a great idea to anyone who remembered the great domestic battles of the 1970s, with the SPLC fighting the Klan and playing a prominent role in the civil rights movement.

Since those proud days, the group founded by Morris Dees has taken a lower-key role in politics. For instance, the SPLC claims its current mission is to "monitor hate groups and other extremists." While it does identify some despicable characters on the fringes of American politics, it also lumps in a number of mainstream Christian conservative groups.

SPLC even runs a "HateWatch" website that demonizes any conservative organization that poses a threat to the left. At one point, HateWatch had 917 organizations listed, with genuinely hateful groups like the Aryan Nation listed right alongside mainstream conservative organizations such as the Family Research Council. All 917 of these groups were being actively associated with genuine extremists, such as the KKK. Branded with labels such as:

Sexist…Intolerant…Xenophobic…Homophobic…Islamophobic…Racist…Bigoted…Anti-Semitic…Misogynist…even Heteropaternalistic now, a newer addition to their anti-hating lexicon (which itself sounds rather hateful to us, just saying).

This hateful branding done in the name of hate watching is about more than words. It's a reversal of the old sticks-and-stones adage—words can and do hurt—and the wounds can run deeper and resist the medication of after-the-fact, meaningless apologies. The HateWatch list has also become the inspiration of radical leftists, who used the HateWatch map to plan out a shooting at the headquarters of the Family Research Council in Washington, D.C., a plan which would have resulted in dozens of deaths if not for a heroic security guard.[58]

The SPLC labeled Maajid Nawaz, a prominent British media personality and founder of Quilliam, as an "anti-Muslim extremist" based on his criticism of specific Islamic groups and practices. It took a court order and a $3.4 million defamation settlement for the SPLC to admit they were wrong, and that Nawaz "condemns both anti-Muslim bigotry and Islamist extremism."[59]

The SPLC also smeared the Alliance Defending Freedom with the "hate group" label. This Scottsdale, Arizona-based legal-advocacy group is well-respected and has won nine cases before the Supreme Court in the past seven years.[60] Its official position is a rather traditional American one:

[We seek] to cultivate a society that is typified by the free exchange of ideas and respect and tolerance for those with different views. These uniquely American and constitutional principles are essential in a diverse society like ours. They enable us to peacefully co-exist with each another. They are the best way to ensure human flourishing.[61]

And then to be branded as a hate group?

Former Attorney General Jeff Sessions spoke before the Alliance Defending Freedom and explained what he believes is the SPLC's rationale:

[They] used this designation as a weapon, and they have wielded it against conservative organizations that refuse to accept their orthodoxy and choose instead to speak their conscience… They use it to bully and intimidate groups like yours, which fight for the religious freedom, the civil rights, and the constitutional rights of the American people.[62]

So far beyond the pale the SPLC has wandered, President Obama's FBI removed the SPLC as a resource on its hate crime webpage. The Pentagon under President Trump dropped the SPLC as a resource in 2017. A coalition of forty-seven conservative leaders, including Protestants, Catholics, Jews, and Muslims, wrote an open letter in 2018 putting the SPLC's prejudices on display.

Through it all, former SPLC spokesman Mark Potok has maintained: "Our aim in life is to destroy these groups, completely destroy them" and that their definition of a hate group has "nothing to do with criminality, or violence…it's strictly ideological."[63]

And this is the organization Facebook is trusting for impartial advice on what kinds of speech should be allowed on its platform?

Worth repeating…without the question mark.

This is the organization Facebook is trusting for impartial advice on what kinds of speech should be allowed on its platform.

Neither waving the white flag on free speech, nor harassing political opponents into silence, have a place in our society. The Constitution is clear on both counts: Free speech is a right, and smothering opposing opinions is not part of the American DNA. Yet some have given up on these core American values. By indiscriminately throwing around its "hate group" labels, the SPLC smothers civil discourse and lowers the quality of civil conversation to the juvenile level now common at US universities among faculty, administration, and students. One has to look (or listen) no further for evidence of this paradigm shift than to the troubling words of Georgetown University law professor Louis Michael Seidman:

> When I was younger, I had more of the standard liberal view of civil liberties. And I've gradually changed my mind about it. What I have come to see is that it's a mistake to think of free speech as an effective means to accomplish a more just society.[64]

So where does Seidman's revisionist thinking lead us, short of anarchy? Would he and his minions in Menlo Park, CA take us back to some dark time before the Magna Carta? The question sounds foolish on its face, unless it's not.

The Algorithmic Art of Skewing Political Searches

Google is responsible for about 75 percent of search engine traffic in the US, and nearly 90 percent on mobile devices.[65] So it matters that in July 2016, with the presidential campaign in full swing, anybody who went to the Google search function and typed in "presidential candidates" would have seen the top feature bar populated with just three names:

Hillary Clinton, Bernie Sanders, and Jill Stein (Green Party).

No mention of Donald Trump or Gary Johnson (Libertarian Party).[66]

This was in July 2016, and the top candidates were too well-known for this to have been an honest error. When alerted to the omission, Google claimed it was a result of a "technical bug" and they soon "fixed it." But when we've seen so many similar bugs, it's hard not to question whether, in Google's eyes, this was a random bug or a well-thought-out feature.

A month earlier, *MarketWatch* had called attention to Google's autocomplete tool.[67] That's the dropdown menu you see when doing a search, and it's Google's AI figuring out what you are most likely searching for with each added letter, based on the billions of searches others have made before you. It's a work of genius and thought to be a completely autonomous tool. But is it, always?

MarketWatch went on to show how Google's autocomplete "tilted in favor of Hillary Clinton" though the company loudly protested to the contrary. The respected financial (and nonpolitical) journal showed a video from *SourceFed* of someone typing "Hillary Clinton cri" into Yahoo and Bing and their autocomplete tool suggesting phrases that linked her to "crime." Similarly, typing "Hillary Clinton ind" on Yahoo and Bing linked her to a possible indictment from mishandling her email records.

But when those identical searches were then performed on Google, the autocomplete gave very different results. Typing "Hillary Clinton cri" gave suggestions for Hillary Clinton crime reform and Hillary Clinton crisis. Similarly, typing "Hillary Clinton ind" brought up suggestions on Hillary Clinton and Indiana, independents, and India, and not indictment.

The *MarketWatch* reporter duplicated the search strings shown in the *SourceFed* video—in the name of good journalism—and obtained the identical results on all three search sites.

Of course, Google denied tampering with the search results, despite a long history of doing just that (which we'll get to). In this instance, they insisted that the autocomplete tool could be influenced by many factors, including the latest queries from the user base, the individual user's own previous searches, and filtering for certain kinds of inappropriate content. And certainly, the past searching habits of individuals on Google, Yahoo, and Bing are going to be somewhat different. But not that markedly different, since search has become a known utility and people don't actively say, why, I think I'll use Bing today instead of Google; they use whatever is in front of them.

So, the Google search results that shined favorably on the Democratic nominee for president were not likely an anomaly. They were more likely an extension of policy and a reflection of the goals of senior management—including Chairman Eric Schmidt—who worked for the Clinton campaign at the time.

These search results could even be considered an "unreported" in-kind contribution to the Clinton campaign under the Federal Election Commission's campaign finance laws.

Moving beyond the presidential race and deeper into the political line-up, the story remained the same: search results biased to the left. Even the liberal *Slate* confirmed it (though with some caveats). They did a crowdsourced study on the last election and found that Democrats fared better than Republicans in the critical first page (top ten search results).[68] Democrats had an average of seven favorable search results in the top ten, whereas GOP candidates had under six—a 40 percent differential that had to have an electoral impact.

As we said, you can go as far back as 2010 to see a developing pattern of "skewed" results at Google. Despite the search giant's early pledge to be a neutral provider of search results, more than one researcher proved that Google was hard-coding bias into their supposedly algorithmic search results, favoring its own products over those of competitors.[69]

For example, the FTC staff noted that Google presented results from its flight-search tool ahead of other travel sites, even though Google offered fewer flight options.

Google's shopping results were ranked above rival comparison-shopping engines, even though users didn't click on them at the same rate. The FTC staff found Google guilty of promoting its own services over rivals, crushing sites such as Yelp and TripAdvisor. In a slapdown of Google that took 160 pages, the FTC concluded Google's "conduct has resulted—and will result—in real harm to consumers and to innovation in the online search and advertising markets."[70]

Yet instead of taking any meaningful action, the Obama-appointed FTC commissioners, who were aware of President Obama's affinity for and allegiance to Google, voted to end the investigation if Google would agree to voluntarily get their house in order.

Google was happy to pretend to comply.

As a private enterprise that built the search platform, Google can certainly decide what happens on that platform. They can even be hypocritical about it. Until, that is, they grow so large they control up to 90 percent of what people see online. Then the calculus changes.

Then Google becomes a clear and present danger to the republic—because they have a monopoly over our most precious asset—information—and they are abusing it.

If Google is unwilling or unable to shoulder such an important responsibility, then a democratic nation cannot allow them to continue being a monopoly.

Like Facebook, Google has entered into several partnerships with "fact checking" companies in an effort to combat fake news. And like with Facebook, you can only hope that Google has selected fact checkers who are as impartial and unbiased as humanly possible. It's no easy task.

The two top fact checkers selected by Google are ProPublica and Pitch Interactive, and the first hint of trouble came when they created a "Hate News Index" to be updated daily with cases of "hate" wherever they are found. We know there's trouble because of all the ways leftists have conflated conservatism and hate in recent years.

ProPublica is a fine organization—shining the media spotlight on hateful and violent acts around the country with a dogged persistence that is admirable. But when it comes to political slant, ProPublica is not exactly unbiased. The Media Bias/Fact Check blog, which grades the point of view of 1,700-plus news sources and comes closest to anyone at ranking without judging, has slotted ProPublica in their "Left-Center Bias" category.[71]

So this is the lens through which Google makes decisions about the search results we see.

And then continuing down this slippery slope of bias, both Google and ProPublica seek out fact-checking advice from the *New York Times, BuzzFeed, Univision News,* WNYC, *First Draft, New America Media, The Advocate,* and others who are known for their left-wing bias.[72]

Throwing Shade on Political Opponents

If you are doing a search on Google, and you bring up a story in the search results from a media outlet that is politically right-leaning, Google posts a "warning" from a left-leaning media outlet. If you reverse the process, no such warning. What's more, the warning you see is not required to get its own facts straight. It is only required, it would appear, to throw shade on the conservative reporting.

An example of this unbecoming and again discriminatory activity was experienced by *The Daily Caller,* a conservative news organization with first-rate credentials. At the time, it was led by the accomplished Tucker Carlson and Neil Patel. If you searched for *The Daily Caller* at one point, you got its page along with Google's own commentary, which included:

1. **A blurb on the news outlet**
2. **Stories it usually writes about**
3. **Stories where Google's fact checkers disputed the facts**

The presumption, unstated but understood, is that conservatives lie all the time, and liberals tell the truth all the time…so Google has to intervene. It's absurdist theater, like blind actors playing to a deaf audience. Google acts the part of George Orwell's Ministry of Truth, which, you'll recall, was responsible for falsifying historical events when necessary.

When Google (a technology company), gets to "fact check" *The Daily Caller* (a news outlet), all kinds of things have gone wrong.

In its own reporting on Google's editorial defamation, *The Daily Caller* detailed how Google's fact-checking tactics went beyond the narrow partisan sentiments of its leaders. It also asserted that *The Daily Caller* made claims they demonstrably never made.

In their zeal to defame conservative media outlets, Google doesn't even appear to care if they, or their hired fact checkers, get their own facts straight. They are more interested in providing a misleading summary of the article. In the example above, Google made it look like a transgender woman raped a young girl because of legislation allowing that to happen. But the original story never made any mention of any such bill being passed. The story was merely a recounting of an article that was first reported by a local news outlet.

Folks at *The Federalist* have been on the receiving end of similar fact-re-arranging from Google. In 2016, writer David Harsanyi penned a story asking why Donald Trump's sexual shenanigans were a problem to liberals when Bill Clinton's were not.[73] It was a well-cited story and flattering to neither man. It mentioned that a "woman, Eileen Wellstone, claimed Clinton raped her while he was at Oxford University in the late 1960s." When you saw the Google search results for this story, Google's fact checker Snopes said the story was false. It was false, Snopes said, because it claimed Clinton was expelled from Oxford for the rape. Thing is, Harsanyi's article never even mentioned Clinton being expelled. Snopes just made it up, apparently.

Months later, Google did apologize and corrected the search results. But *The Federalist's* takeaway was: Policy at the search giant is to "hurt conservatives very publicly, apologize in private later."

Writing for *American Thinker*, Monica Showalter summarized the situation spot-on:

> *Google has taken to throwing shade almost exclusively on conservative websites through its search engine mechanism, using a sort of "fact-checking" system to discredit certain news-providers so no one will want to click on them. Kind of an odd thing for a search engine company to do, given that its business is built around gaining clicks. But it's not the first time the Silicon Valley giant has been accused of disguised censorship against conservative news outlets under the guise of the war on fake news.[74]*

A review of both conservative and liberal news outlets shows that only conservative outlets are subjected to Google's "reviewed claims" process. Left-leaning and even radically leftist sites like *Vox, ThinkProgress, Slate, The Huffington Post, Daily Kos, Salon, Vice,* and *Mother Jones* are spared.

For example, *ABC News* suspended reporter Brian Ross after he filed a fake news story on Michael Flynn.[75] That fake news story was indexed by Google and included in the search results. But did Google post a "reviewed claim" on the *ABC News* blurb? Of course not.

After conservatives lashed into Google for their fact-checking reign of errors, Google suspended the practice. But as we'll see, they had several more tricks up their sleeve for squelching ideas that don't align with their own narrow worldview.

Google Gets Religion—Just Not Christianity

All kinds of digital assistants have made their way onto the countertops in America's homes. These AI-driven machines can make life easier with just a voice command. Playing your music. Controlling the temperature and sunlight. Creating on-the-fly shopping lists. All just by listening to what you say.

We'll talk in the third section about the privacy aspect of these devices. Our concern here is how these assistants, such as Google Home, were originally programmed by humans and what that means. Because these programmers are the same people who sign "Rot in Hell" petitions against Donald Trump and think of Jesus as some, well…here's a story that tells itself—as reported in *Breitbart News*:

> Google's Home A.I. assistant refused to define Jesus Christ in an experiment recorded by television producer David Sams, but happily defined Muhammad and Buddha. When asked who Jesus Christ is, the Google Home replied, "Sorry, I'm not sure how to help" or "My apologies, I don't understand."[76]

The producer next asked Google if it knew who he, David Sams, is. No problem with that question. His bio came right up. Sams concluded: "Google did not know who Jesus Christ was, and Google did not know who God was… It's almost like Google has taken Jesus and God out of smart audio."

A more recent query about Jesus by another reporter using Google Home gave no reply and instead sent the reporter to a Jehovah's Witness website.[77] What, were some software engineers sitting in their cubicles at Google one day thinking up ways to mess with people they don't like? Doing it…because they can? It seems far-fetched, right? But instead of conjecturing about it, Sams reached out to Google for comment.

Their spokesperson claimed that the Google Home response (and lack of response, as well) was an effort to ensure respect, not disrespect. Huh? So acting like Jesus didn't exist, but acting like Muhammad and Buddha did exist, were signs of respect? Or conversely, would a mention of Muhammad or Allah be somehow misconstrued as disrespectful to Muslims? It's like the spokesperson wasn't even trying to come up with an excuse.

We readily acknowledge that programming a digital assistant to answer a gazillion questions is no easy feat. But when you examine Google Home's handling of the historical Christ vis-à-vis the historical Muhammad and Buddha, you have to wonder if Google engineers were playing dice with the search universe. Playing God themselves and laughing at the rest of us.

Case After Case of Censorship Piling Up

Here is just the beginning of a long list of top conservative websites that have been blacklisted and targeted by Google, YouTube, and Facebook.[78] And as you can see from the first name on the list, the social platforms are gunning all the way to the top.

President Trump

As mentioned earlier, Facebook's February 2018 algorithm update caused Trump's engagement on Facebook posts to plummet a whopping 45 percent—dropping from 4.8 million engagements to 2.6 million in less than a month.[79] In the same period, according to a *Breitbart* analysis, Senators Elizabeth Warren and Bernie Sanders did not appear to suffer any comparable decline in Facebook engagement.

The Declaration of Independence

Somewhere in paragraphs 27-31 of the Declaration of Independence, Thomas Jefferson wrote something that Facebook considers "hate speech." This is not some satirical intermission, dear reader; this is serious. Quite serious, as a small Texas publication, *The Vindicator*, learned.

For a July Fourth celebration, *The Vindicator* challenged its Facebook followers to carefully read the Declaration of Independence. They broke it into chunks—one small bite over the course of twelve mornings. The first nine bites flew through Facebook without a hitch. But on the tenth, paragraphs 27-31 of America's core founding document, they were banned. Facebook notified *The Vindicator* that content in paragraphs 27-31 "goes against our standards on hate speech."[80]

Apparently the reference in those paragraphs to "Indian savages" tripped up the Facebook algorithms. Understandable, sure, except that this points out how fallible and fraught it can be to rely on algorithms for upwards of 90 percent of information people ingest.

After being called out by critics, or rather laughed out, Facebook did reinstate the post.

Natural News Channel

The *Natural News Channel* is a prominent health and wellness website with tens of thousands of YouTube viewers. In one fell swoop, YouTube stepped in and wiped out the channel's 1,700 videos covering everything from nutrition, natural medicine, history, science, and current events.

Lloyd Marcus

Perhaps on a roll, Facebook next banned a music video titled "We Are Americans" from Lloyd Marcus—a singer/songwriter. In his words:

Facebook rejected my Pro-America Christian July 4th music video. My wife Mary tried to purchase a boost ad on Facebook to promote the July 4th release of my "We Are Americans" music video. The video was rejected for "political content."

It must have been the "We Are Americans" title that so offended the Facebook censors. The video was later allowed, after Independence Day had come and gone.

Wes Cook Band

The popular Nashville-based country music group with frontman Wes Cook was shocked to learn their latest song, "I Stand for the Flag," was stricken from Facebook. The social platform initially approved and then rejected and later reapproved the band's request to buy ads to promote their song. Apparently, the lyrics were causing a lot of confusion at Facebook HQ.

So what was so bad about Wes Cook's lyrics? We invite you to visit his website to see for yourself. You'll find the lyrics are saying, basically, that Americans of all races, creeds, and political affiliations can come together and support the flag of our nation with pride. There is absolutely nothing in the lyrics that could cause Facebook's censors indigestion. Just the opposite, it's a song about inclusion—something you'd think would please the censors in the Bay Area. As Cook told Fox News:

> Our song bleeds unity and love of country. It doesn't see race, color, religion, or political affiliation…"I Stand for the Flag" means I am thankful for the freedoms and opportunities this country gives all citizens and shows how dependent we all are on each other for the success of our individual American Dream. I believe patriotism is not political.[81]

No finer words. It is good to stand for the basic principles of freedom and liberty, and work to improve on the execution of these principles constantly. But with Facebook and Google as gatekeepers, the ideals of freedom and liberty have been supplanted by division and censorship.

InfoWars

Another popular site is *InfoWars*—run by the well-known Alex Jones, who often has some very controversial claims, such as the one about Democrats starting a Second Civil War on Independence Day. All of the big-name social platforms, including YouTube, Facebook, Apple, and Twitter have banned him. They have not given specifics other than Mr. Jones has "violated their community standards," which is the common reason given to many other conservatives that have been banned from those platforms.

If there is any truth in the conspiracy world, it's that conspirators come in every stripe—from the political right and left. This has always been true, and we presume, always will be. The bigger the megaphone they are given, the louder they will be perceived as shouting. That's the often harsh truth of the platforms these technologists have created.

But banning just the controversial guys on the right, while letting the same type of guys on the left run unchecked, betrays the trust in a neutral platform of ideas. It's not supposed to be about the goodness or the badness of ideas on a supposedly neutral platform. Up to the point where harm is intended, every idea is free in America to be judged by the marketplace, not purged selectively by the platforms.

Luke Rudkowski

Luke Rudkowski, an independent journalist and activist who runs *WeAre-Change*, told *The Daily Caller* that hundreds of his videos were demonetized in a single day, effectively killing his ability to earn a living on YouTube. Rudkowski says,

> Having had 660 of my videos demonetized in one day left me a little stunned since this is the core for my income but left me with the impression that this was done on purpose… After dealing with all the repeated issues with YouTube, it is clear that this is a campaign to de-incentivize any critical thinkers and anti-authoritarians from their platform.[82]

Right Wing News

Thanks to its popularity on Facebook, *Right Wing News* grew in leaps and bounds, reaching 133 million online viewers in 2015. With 3.6 million Facebook "likes," this popular landing place thrived…until after the 2016 election, when Facebook began hiding views of its pages. Right Wing News continues today because of the tenacity of owner John Hawkins, an entrepreneur with deep roots in the conservative community. But unlike his counterparts on the left, he has no real presence on Facebook now.

LifeSiteNews

In the wake of Ireland's May 2018 abortion referendum, Facebook began denying the efforts of *LifeSiteNews* to advertise on its platform. Facebook says the photos of things like ultrasounds, the bellies of pregnant women, fetuses, and baby feet used in pro-life ads are "too offensive" for their viewers and they wish to remain "neutral."[83]

If Facebook was committed to remaining neutral in the abortion debate, as they claim, why in the wake of the abortion referendum did they permit a torrent of ads from Planned Parenthood to run on their platform?

What's more, Planned Parenthood has run ads featuring images of condoms on bananas with copy declaring "Sex is hot—bad sex ed is not," which may well be a violation of Facebook's rules for underage audiences.

Planned Parenthood has also run expressly partisan ads, even targeting politicians by name, accusing them of disrupting access to health care—which itself could be found false and defamatory.

So Facebook is clearly not neutral, not nonpolitical, and not concerned that their platform is a conduit for potentially libelous statement.

Gab

Apple and Google combined control 98 percent of the mobile-phone operating systems, so any company wishing to launch a mobile app must run through them. A new company named Gab tried just that—but was banned.

How the "hate speech" label got slapped on Gab is a mystery to its co-founders, Andrew Torba and Ekrem Büyükkaya. One is a Turkish Kurd, the other a Muslim. They are concerned about speech repression because of the Turkish government's growing suppression of free speech and imprisonment of journalists. These are people on a mission—you could even call them the good guys.

But Apple and Google don't.

The Gab app is like Twitter in how it works. It features a free-speech policy modeled on the First Amendment itself. Enough people were attracted to Gab's value proposition that they quickly built up to 200,000-plus users. But the app, tailored to run on Apple's iOS operating system, couldn't get an approval. And Google flagged them, citing "hate speech" as its reason for excluding them from the platform.

Google offered nothing in the way of a helpful clarification of what they meant by hate speech, just the vague label of "hate speech" that they could hide behind. Google also didn't engage with Gab in a back-and-forth discussion that might resolve whatever vague problem Google was having. Just a flat no. Like the "your eyes are red, you must be the devil" defense, there was no logic in Google's actions, or in Apple's. Only the hammer of restraint of trade, swung down with extra delight, no doubt, because of the political beliefs of the site it was leveling, and destroying.

Meanwhile, apropos of the labeling of "hate speech," you can still go to Google and successfully search to find the latest ISIS recruitment pictures, graphic child pornography, and live streaming of murder and torture. The sewage runs deep. But you can't find Gab.

If anything, Google seeks to be known as an authority on the dissemination of hate speech—the kind they like, that is. In 2016, the search giant brought onto the payroll Chris Poole. He had been the founder of the notorious website 4chan, known for revolting content that included child pornography—for which arrests have been made.[84] Yes, Google hired a filthmonger and allowed his apps in their app store. But not Gab's.

Apple and Google are instead using their dominant market positions to restrain trade, squelch competition, and revoke the First Amendment on mobile phones. So, for all the online activity that takes place on apps—and it's now the majority of online activity—the First Amendment has been cleaved from the Constitution, and turned into whatever Apple's and Google's Terms & Conditions say it is.

There is no sign this will change unless action is taken at the highest level.

Independent Journal Review

Another big conservative website driven by Facebook marketing, *Independent Journal Review*, took a massive hit when Facebook arbitrarily blacklisted them.

SarahPalin.com

With over four million fans, Sarah Palin obviously had a popular page on Facebook. But then Facebook began blocking traffic to the website.

The Gateway Pundit

Ranked in the top five most influential conservative news outlets during the 2016 election, *The Gateway Pundit* had for years received about a third of their traffic from Facebook. Then suddenly Facebook blocked all traffic links to the website—without explanation. The site was blacklisted, despite having paid

Facebook for ad promotion…moving Facebook's action into the realm of possible fraud.

The Gateway Pundit has also been shadow banned by Google.

Rightside Broadcasting

The YouTube channel of *Rightside Broadcasting* had millions of views before the 2016 election. After the election, YouTube shadow banned all their videos. The video-hosting platform classified *Rightside's* videos of a Trump rally as "hate speech."

In addition to shadow banning, YouTube demonetized hundreds of *Rightside* videos—costing the broadcaster an estimated 95 percent of its revenues.

Prager Report

This leading conservative educational site, PragerU, with an audience in the millions, came to work one day to find YouTube censoring one video after another for no other reason than their political leanings. Among the banned topics were abortion and Islam—topics that left-wing sources were still able to post about without repercussion. So PragerU has taken YouTube to court. (More on PragerU's fight to come).

Western Journalism Center

During the 2016 election, *Western Journalism Center* (now known as the *Western Journal*) was averaging an impressive six million page views a day (third-party "Quantcast" stats). Then it, too, was blacklisted and lost as much as 80 percent of the previous levels of traffic. (More on *Western Journal's* fight to come).

Daniel Sulzbach

Known for his blistering critiques of atheism and video games, Daniel Sulzbach says he doesn't understand why his YouTube videos are now being flagged and demonetized: "I don't make any crazy radical videos…I hardly even do videos regarding feminism, social justice, etc., anymore."[85]

So it would appear that on behalf of feminism and social justice, YouTube was banning Sulzbach's views on feminism and social justice. Not only that, but retroactively.

In YouTube's view, it would seem that the promotion of social justice requires eliminating any opinion on social justice that doesn't comport with its own. Like Justice holding the scales, except now her blindfold is loosened so she can peek out of her left eye.

Young Cons

This popular conservative news site aimed at millennials had millions of daily readers during the 2016 election. When Facebook next slapped them down, they went on life support and now struggle to stay afloat.[86]

Daniel Harris

Edgy, profane, provocative, using the stage name Razorfist, Daniel Harris is not everyone's taste. He's like a Lenny Bruce…on the right. And when he came out in support of Donald Trump, all of the videos he created were abruptly demonetized. Harris believes there is a white list for leftists and a blacklist for rightists. Leftists can say whatever they please, rightists get penalized. In his view:

> Google's going to need to explain to me why John Oliver can engage in weekly invective punctuated by a hail of profane epithets, skew it leftward, and still have ads for Pampers and pimple cream adorn the margins of his unwatchable videos (though I cannot).[87]

Diamond and Silk

These famous pro-Trump YouTube sensations had millions of viewers only to discover in August 2017 that YouTube had demonetized 95 percent of their videos. In April 2018, the sisters testified before Congress regarding their experience with Facebook bias and censorship.

Mark Dice

He's a conservative video blogger who, in interviews with *The Daily Caller*, has linked YouTube's censoring to the video hosting sites' own internal anger at the rising popularity of conservative authors—which many believe had historically lagged their liberal counterparts in messaging talent:

> I think YouTube is furious that so many conservative channels have gotten so popular in the last year, and they don't want us to be able to work full-time doing what we're doing because our message is at odds with almost everything that Google and YouTube's leadership stands for.[88]

Like the others listed here, Dice has seen his revenues drop from YouTube's censoring and selective demonetizing.

Pamela Geller

Well-known for speaking out against radical Islam, activist Pamela Geller has been threatened with beheading multiple times by ISIS. During the 2016 election, she averaged 100,000 daily views on PamelaGeller.com from traffic originating on Facebook. Then suddenly Facebook shut down most of her traffic. Why?

Because terrorist groups and their radical leftist sympathizers were demanding that they shut her down. Whether Facebook buckled under, or felt some affinity with the terrorists' demands, we don't know. But Facebook's actions did cost Geller a serious loss of revenue. This, for someone willing to put her life on the line for our country. In this case, Facebook's actions are both despicable and cowardly.

We wish this list of offenses was a complete one.

But many more conservative websites and video producers have been banned by Facebook, Google, Twitter, and Apple. Some are even afraid to speak publicly for fear of losing any more traffic referrals from the social platforms.

Again, while the activities of these social platforms do not reach the level of punishment meted out by the East German Stasi, there are uncomfortable resemblances…

- Social platforms do control the public information channels just as the Stasi did.
- Social platforms do try to keep citizens in line with 24/7 surveillance techniques.
- Social platforms *cannot* directly throw citizens in irons, of course, the way the Stasi could, but they *can* silence speech and *are* startlingly effective at it.

And that makes the comparison apt.

It's as if the old slogan of "Don't Be Evil" has smudged into "font of evil," and the people behind it have no idea how far they've wandered down the slippery slope of censorship. They certainly don't think of themselves as evil people. They just have at their disposal the best artillery in the modern social wars.

Bias in Google Search Referrals

A good tool for measuring the amount of traffic that crosses a website's threshold is Alexa (yes, Amazon's tool known to webmasters long before it became a listening and ordering device). A researcher named Leo Goldstein used Alexa to run a straight-up comparison of how much traffic Google sent to liberal sites, and how much it sent to conservative sites. He readily admits that Alexa is a tool of approximation, and is subject to error. But his findings, as summarized in the next page's chart, are simply astounding at the 30,000-foot level. Here is his own summary:

> Google Search is found to be biased in favor of left/liberal domains and against conservative domains with a confidence of 95%.... Further, certain hard-Left domains have such a high [percentage of domain traffic, referred by Google Search, net of brand searches] that their standing raises suspicions that they have been hand-picked for prominent placement.[89]

Based on his findings, Goldstein believes a number of conservative sites—specifically *American Thinker, Drudge Report, Power Line Blog, PJ Media,* and *The Gateway Pundit*—have been dark-targeted by Google in an effort to drive down the actual search results they receive. This is his conclusion, based on how popular these sites are, and how relatively little traffic they received from Google in his study period.

This next story takes another look at that artillery, and the limits of countermeasures.

PragerU v. YouTube—Taking the Censors to Court

Dennis Prager is a popular radio host, author, and Jewish scholar. The *Los Angeles Times* has called him "an amazingly gifted man and moralist whose mission in life has been crystallized—'to get people obsessed with what is right and wrong.'" Toastmasters International has called him "one of America's five best speakers." And *Buzz Magazine* said he's "one of the ten most powerful people in Los Angeles...a moral compass." But YouTube called his teachings "inappropriate" for young viewers.

Prager had spent six years building PragerU, an online university featuring videos for educational purposes. These videos present counterarguments to leftist views on America. One day, YouTube just decided to shelve as many as fifty of PragerU's videos, including:

- "The Most Important Question About Abortion"
- "Is Islam a Religion of Peace?"
- "Why America Must Lead"
- "The Ten Commandments: Do Not Murder"
- "The World's Most Persecuted Minority: Christians"

YouTube declared these titles "inappropriate" for young viewers. But how exactly is a video cautioning against murder, for example, inappropriate? If anything, it's the opposite!

Once he got over his initial shock at being throttled for producing quality videos, Prager challenged YouTube to defend its position with some semblance, any semblance, of logic. Prager wrote:

> Our videos are presented by some of the finest minds in the Western world, including four Pulitzer Prize winners, former prime ministers, and professors from the most prestigious universities in America… Watch any one of our videos, and you'll immediately realize that Google/ YouTube censorship is entirely ideologically driven…They are engaging in an arbitrary and capricious use of their "restricted mode" to restrict non-left political thought.

Indeed, the PragerU catalog of videos are narrated in a thoughtful, even tone. No yelling, no profanity, no screeching, no shouting—just a considered perspective. One of the blackballed videos was a discussion with esteemed Harvard Law Professor Alan Dershowitz. Another was among the first to warn about the dangers of e-cigarettes. Another was a discussion with Bret Stephens—a *New York Times* columnist. Clearly educational tools of value to many people.

Like so much of the censorship we've seen, YouTube's actions appear to be all about politics. But it's also about a larger thing. It goes to the heart of the Constitution—to the First Amendment right to free speech and whether that right will endure or be smothered in arbitrary and capricious ways because new technologies make it possible. Which is why Prager went to court against YouTube, filing suit in October 2017. In the complaint, Prager contended:

- YouTube's policies are an obvious pretext to justify illegally discriminating against PragerU because of its conservative political perspective and identity.

- YouTube's content policies are vague and lack "objective criteria," allowing the site to get away with cracking down on conservative voices.
- Singling out conservative voices for such treatment is speech discrimination, plain and simple, and by doing so Google violated PragerU's First Amendment rights, engaged in unlawful discrimination under California law, and violated its own terms of use.
- This censorship is profoundly damaging because Google and YouTube own and control the largest forum for public participation in video-based speech in not only California, but the United States, and the world.

PragerU had worked with YouTube for over a year, trying to remove its videos from the "restricted list," but all they got were "conflicting, vague and unhelpful answers from the company."[90]

YouTube clearly had a hit list, based on their own double standard, as became clear in this *Investor's Business Daily* exposé:

> YouTube labeled a PragerU video titled "Why America Must Lead" as inappropriate, but not a video by Sen. John McCain titled…"Why America Must Lead." It even found instances where the exact same video was restricted when it appeared under the PragerU label, but not when it was posted by someone other than PragerU.

That is blacklisting, it is discrimination, and beyond the ethics of it, that is the very definition of unlawful business practices in this country. A lawyer as learned as former California Governor Pete Wilson explained why the legal case should be cut and dry:

> This is speech discrimination plain and simple, censorship based entirely on unspecified ideological objection to the message or on the perceived identity and political viewpoint of the speaker.[91]

There are more examples of YouTube's discriminatory double standard. *The Washington Examiner* found that PragerU's video "Is America Racist?" has been restricted on YouTube, but the leftist *Vox* video "The Racist History of U.S. Immigration Policy" is available for viewing.[92]

So how does YouTube make these decisions? It's not based on the keywords in the title, obviously. It's not based on the production quality—for PragerU's was of theatrical release quality. So what?

The logical answer seems to be: It's the messenger YouTube seeks to silence.

Prager himself best summed up the position that our country has devolved into: "Google, and their wholly owned company YouTube, apparently believe they can pick and choose who has free speech in this country."[93]

In an ideal marketplace, driven by competitive market forces, there would be other platforms that publishers could use to reach audiences. But it isn't, and there aren't—not anymore.

PragerU's day in court came in March 2018. Presiding US District Judge Lucy Koh threw the lawsuit out, ruling that Google's YouTube did not violate any First Amendment rights in their blacklisting because Google has no obligation to equally apply its services. It can discriminate all it likes, Judge Koh ruled, even as the dominant social platform.

The decision came as no surprise, really. Judge Koh was appointed to the Northern District of California by President Obama on the recommendation of California Senators Barbara Boxer and Dianne Feinstein—the two ladies carrying water for the social platforms so critical to their many election victories. However, Feinstein would later come to realize why the platforms truly need sanctioning, as we discussed earlier.

Marissa Streit, PragerU's CEO, described the ruling as "far from an unexpected setback" and said that further avenues of recourse remain: legally, the highest court; politically, the White House.[94] As of January 2019, PragerU "is now running dual tracks at both the California State Court as well as with the Ninth Circuit Court of Appeals." PragerU says:

> …[T]here is reason to believe certain claims are even stronger in California. Specifically claims relating to YouTube's breach of contract and consumer fraud. They claim to be a public forum for free expression, but they behave instead as a publisher with editorial controls. You cannot have it both ways…. Rest assured, we will not stop fighting to secure all Americans' freedom of speech and expression online.[95]

A Detailed Study of the Impacts of Facebook Censorship

We've been writing about the long list of conservative publishers demoted by the social platforms. We've shown proof of these demotions, and shared beliefs and opinions, as well, because not every organization has been able to prove damage done. Here we are reprinting, in an abridged format, an in-depth analysis conducted by *The Western Journal* that details precisely what damage was done.[96]

From The Western Journal

Liberal publishers have gained about 2 percent more web traffic from Facebook than they were getting prior to Facebook's algorithm changes implemented in February 2018. On the other hand, conservative publishers have lost an average of nearly 14 percent of their traffic from Facebook. This algorithm change, intentional or not, has, in effect, censored conservative viewpoints on the largest social platform in the world. This change has ramifications that, in the short term, are causing conservative publishers to downsize or fold up completely, and in the long term could swing elections in the United States and around the world toward liberal politicians and policies.

Case in Point: New York Post vs. New York Daily News

Two rival publishers in New York City, the *New York Post* and the *New York Daily News*, are similar in their bombastic ways, though opposites in editorial slants. The *Post* leans right, the *Daily News* leans left. Here's an example of how each outlet covered the terrible Stoneman Douglas High School shooting in February 2018:

Left-leaning *New York Daily News:*	Right-leaning *New York Post:*
Florida school shooting survivors hit lawmakers who oppose gun control with ultimatum[97] Received a **24.18 percent** INCREASE in traffic from Facebook in the same time period.	**Florida school shooting suspect is 'troubled' former student obsessed with guns**[98] Received a **11.44 percent** DECREASE in traffic from Facebook in the same time period.

How Facebook Went About Changing Their Role in Society

After Facebook's latest round of skirmishes with lawmakers and consumers, the social platform hired former NBC and CNN honcho Campbell Brown to help right the ship. She was put in charge of Facebook's news partnerships team. During a speech at a conference of technologists (the Recode conference)

Brown basically laid out Facebook's new plan to censor news publishers based on its own internal standards:

> This is not us stepping back from news. This is us changing our relationship with publishers and emphasizing something that Facebook has never done before: It's having a point of view, and it's leaning into quality news… We are, for the first time in the history of Facebook, taking a step to try to define what 'quality news' looks like and give that a boost.[99]

Hearing the speech, it sounded to many people like Facebook's definition of "quality news" would be news with a liberal slant. But *The Western Journal* sought proof—one way or the other. The *Journal's* research team selected fifty publishers known to receive a large portion of their online traffic from Facebook. These publishers included traditional print or television outlets such as *The Washington Post,* CNN, and Fox News, as well as new media outlets like *Salon, Vox,* and *The Daily Caller.* All political leanings were included.

Each publisher was assigned a number between zero and one hundred supplied by *Media Bias/Fact Check News,* a third-party website that analyzes publishers for political bias and places them on a continuum between "extreme left" and "extreme right."

The monthly Facebook traffic for each of these sources was then established using data from the market intelligence company SimilarWeb. January traffic was compared to February traffic—for a before-and-after snapshot.

Results: Conservative Publishers Negatively Impacted

The twenty-five news organizations on the liberal side of the scale averaged a 1.86 percent boost in traffic from Facebook during this period. The twenty-five news organizations on the conservative side averaged a 13.71 percent decrease in traffic.

After removing the fifteen publishers that receive very little traffic from Facebook, and are thus presumably of little concern to Facebook, the trend became even more clear:

- The twelve "most liberal" sites averaged a boost of only 0.21 percent—in other words, they don't appear to have been affected meaningfully.
- The eleven sites in the middle—which ranged from "left-center" to "east biased" on the Media Bias/Fact Check News scale—saw a sharp increase in traffic of 12.81 percent.

- The twelve most conservative sites lost a significant 27.06 percent of their traffic.

Did Facebook Intentionally Target Conservatives?

- Of the twelve most liberal sites, six saw double-digit decreases in traffic, while four saw double-digit increases and two—the *Washington Post* and *HuffPost*—saw single-digit increases. CNN's traffic increased 43.78 percent.
- Of the eleven sites in the middle of the scale, nine saw traffic increase. Only two—CBS News and *The Atlantic*—saw a traffic decrease.
- Among those eleven, only two—*USA Today* and *The Economist*—can truly be considered centrist according to the Media Bias/Fact Check News scale. Their traffic increased by 23.16 percent and 1.12 percent, respectively.
- Of the twelve most conservative sites, only two benefited from increased Facebook traffic—the *Daily Mail* with 3.51 percent and Fox News with 31.67 percent increases.
- If Fox News was removed from the group of twelve conservative sites shown above, the average drop would be 32.4 percent among the remaining eleven.
- The other ten saw decreases ranging from a modest 3.13 percent at Breitbart to a crushing 76.4 percent decline at *Independent Journal Review*.
- Rare, a conservative publication owned by Cox Media Group, experienced a 68.7 percent drop in traffic after the algorithm change, and was forced to shutter its doors.[100]

Facebook's Official Response to the Charges

It is always possible that the benefit to liberal sites and the harm to conservative sites is unintentional, a side effect of Facebook's well-known "move fast, break things" attitude. But given Facebook's comments acknowledging that Facebook will have a point of view going forward, it appears that Facebook has made an intentional break with its longstanding goal of being a neutral platform. Brown, head of partnerships at Facebook, made that clear to the technologists gathered at that Recode conference we referenced earlier:

> How this manifests in the coming months is not totally clear to us right now...These are conversations we've just started having with a lot of publishers. But in terms of us taking a big step in that direction, I think, yes, I think this is, I think this is us having a very clear point of view.[101]

That point of view was first unleashed on conservative publishers, then on conservative lawmakers. Once again, *The Western Journal* studied it...[102]

US Congress Members Selectively Targeted by Facebook

Following Facebook's January 2018 algorithm changes, pages associated with members of both major parties saw a significant decrease in interactions with readers. But the pages of Republican members of the House and Senate were impacted measurably more than those of their Democratic counterparts.[103]

The Western Journal study of the interaction rates on these political pages found: GOP congressional pages decreased 37 percent, Democratic congressional pages decreased just 27 percent.

The Western Journal downloaded months of pre- and post-algorithm change data through CrowdTangle, a social monitoring platform owned by Facebook. The period of study was August 2017 to June 2018. (January 2018 data was omitted because the algorithm change was made partway through the month.)

The Western Journal looked at total interactions (reactions, comments, shares of a post) as well as interaction rates (average interactions divided by the number of page followers). Regardless of a change in the number of posts or followers, the interaction rate on a page should remain similar from month to month, all else being equal.

According to this analysis, the interaction rate on Republican Facebook congressional pages decreased 37 percent, compared to a change of just over 27 percent on Democratic pages.

Moreover, total interactions on Republican congressional members' Facebook pages decreased 34 percent, while total interactions on Democratic congressional members' Facebook pages decreased only 18 percent.

While interaction rates fell across the board after Facebook's algorithm change, Republican senators and House members saw a more significant drop in the interaction rates on their Facebook pages than did their Democratic counterparts.

Among senators, the interaction rate on Republican pages decreased 44.37 percent, while Democrats saw only a 31.97 percent decrease.

In the House of Representatives, the interaction rate on Republicans' Facebook pages decreased 31.58 percent while the interaction rate on Democrats' pages decreased 25.10 percent.

Total video interactions on Republican Facebook pages decreased 49 percent while total video interactions on Democratic pages decreased 28 percent.[104]

The interaction rate on Republican members' videos declined 46 percent, while the rate on Democratic members' videos decreased 33 percent.

It should be noted that, since the number of videos posted during the periods in question was relatively small, a wider variation in these numbers is to be expected.

Facebook Users Saw Fewer Photos of Targeted Congress Members

Photo interactions on Republican members' pages decreased by 24 percent, with a 35 percent decrease in interaction rate.

Photo interactions on Democratic pages declined by only 6 percent, with a 26 percent decrease in photo interaction rate.

Facebook Users Saw Fewer Links of Targeted Congress Members

The link interaction rate decreased on Republican Facebook pages 32 percent, while the Democratic pages had only a 17 percent decrease.

Asked to comment on these findings by *The Western Journal*, a Facebook spokesman said:

> We updated News Feed to help people meaningfully connect with friends and family first. As News Feed prioritizes posts from friends, it means public pages of all types have experienced declines, regardless of political perspective. In fact, while our own analysis shows decreases across the board for Republican and Democratic member Facebook pages, there have been large spikes driven by conversation about news events, such as tax cuts last December and the immigration debate in June.

It is true that if December 2017 and June 2018 were removed—as these months had far more interactions than other months in the data sample—the difference in interactions and interaction rates would be much less. *The Western Journal* left those months in the analysis because (1) rises and falls in political engagement are a normal occurrence, and (2) including all months in an analysis, instead of a few handpicked months as the Facebook spokesman did, presents a more accurate picture of what really happened.

Conclusions:

- The ability of elected officials in Congress to communicate with constituents through Facebook was markedly reduced by the January 2018 algorithm change.

- Facebook's lack of data transparency makes it impossible for any outsider—from The Western *Journal* to government regulators—to defend Facebook's internal processes as unbiased, make a credible accusation of intentional bias, or make any sort of defensible statement in between.

Google Crushing Competition, Restraining Trade

Crushing a Better Search Engine

It was a magic moment for Adam and Shivaun Raff in early 2006. They had just created MatchMate, a dating app that let people mix variables in their searches (what's known as parameterization). It was at the time a clever improvement on the dominant search methodologies. So the Raffs applied for a couple of patents. Even better, they learned that their search technology could work on almost any data set. That is, it could search for the lowest-cost airline tickets or the highest-paying jobs. It could also handle apples-and-oranges questions, such as "What's the cheapest flight between London and Las Vegas if I'm trying to choose between business class or leaving after 3 p.m.?"[105]

The couple knew they were onto something even bigger than dating. So they launched Foundem.com—a search engine for finding cheap online prices. Their first two days as a live site saw a rush of traffic from people doing product price comparisons. Then suddenly it was honeymoon over, traffic stopped.

Alarmed, the Raffs began running diagnostics. They soon discovered that Foundem had slipped off the first page of Google's search results and could only be found dozens of pages in. On other search engines like Yahoo, Foundem still ranked high. But on Google, Foundem had been disappeared.

Turns out, Google had gone to war. Once Foundem started getting traction—because it was a superior product—Google went ahead and manually elevated its own price comparison results to the top of most every page of search results, overriding the organic results of their own search algorithm.

Yes, Google had gone to war, and Foundem was the first casualty, just as TheTeaParty.net would later be with Facebook…in a pattern that would become a familiar one.

A company would develop a product that posed a threat to Google's business. Google would step in, and saying nothing publicly, manually lower them in the search results.

But that's not all they would do.

Crushing a Better Navigation System

Take the case of Skyhook Wireless. They had developed a navigation system that cellphone makers like Samsung and Motorola found very appealing. Deals were signed. Skyhook saw a bright future ahead. Google wanted in on that business, but according to their own engineers, the Google solution didn't measure up to Skyhook's.

Oh well!

Or not oh well, because days later, according to court records reported by *The New York Times*, a senior Google official made it very clear to Samsung and Motorola that if they didn't deal out Skyhook in favor of Google, they'd risk seeing unexplained shipping delays in their phones.[106] Google later denied attempting any such blackmail. But Samsung and Motorola did tear up their Skyhook contracts soon thereafter.

Skyhook later sued Google and won a $90 million patent infringement settlement—only then did all this skullduggery became public. But Skyhook had been ruined by then, and were forced to unload their remaining assets at a loss.

Google has succeeded where Genghis Khan, communism and Esperanto all failed: It dominates the globe.

—CHARLES DUHIGG, PULITZER PRIZE-WINNING COLUMNIST AT
THE NEW YORK TIMES AND AUTHOR OF THE BEST-SELLERS *THE
POWER OF HABIT* AND *SMARTER FASTER BETTER*[107]

Crushing Better Online Reviews

Yelp is one of the best-respected brands in the online world, which is why Google tried to buy them and, failing that, buried them.

Yelp's website—with millions of user-generated reviews of local salons, pubs, auto-body shops, you name it—was grabbing a third of all online searches. If you wanted the "Best Barber Shop in City Name," you trusted Yelp, and you found their reviews in the top pages of Google for just about all relevant keywords.

So what did Google do? They commandeered Yelp's content. They just scraped up Yelp's reviews and embedded them into Google's own search results. That meant many users no longer visited Yelp's website. To users, it may not have seemed like any big deal. Why should they care how they accessed the Yelp reviews, as long as they could?

Well, you know Yelp cared. They had spent time and treasure to organize a national corps of reviewers into a valuable business service of great distinction. Then one day a company was allowed to lift all their intellectual property. It was—and there was a term for it, back in the day—corporate theft.[108]

Same thing happened at sites like TripAdvisor and Citysearch, as well as the Getty Images search engine. The companies are all shells of themselves now, victims of the kind of monopoly power the government used to outlaw.

Crushing Conservative Social Networks

Rarely is Google more venomous than when they face a potential competitive threat from a website that is also backed by conservatives and has the audacity to publicly criticize the Big Tech Tyrant.

Such is the case with Minds.com—a fascinating new "crypto" social network where users get paid in cryptocurrency for their contributions to the community. It's very cutting edge. But that's not why CEO Bill Ottman found himself on Fox News talking to Tucker Carlson. That happened because of what Google did to Minds.com:

> We got banned from [Google's] advertising platform...probably caught up in their out-of-control algorithms which basically blanket-ban companies based on certain keywords with no real rationality...big tech companies are now a far bigger threat to civil liberties than the federal government ever was.[109]

This banning meant Minds.com was shut out of Google's ad platform, which can be death to a new company because they're unable to reach new viewers. Minds.com is hoping instead to develop their own ad platform and end-run Google. Their aim is to be completely transparent with open source code, to ensure true freedom of speech and non-censorship. But the company's battle will be a difficult one—like David trying to find a needle in a haystack infested with Goliath needles.

And so in recapping, we've dragged out into the light the four ways Google pushes its agenda and perspective onto others—ranging from bad-faith partisanship to provable restraint of trade:

1. **Skews political searches toward Google's favored liberal candidates and ideals.**

2. **Falsely scrutinizes conservative voices and perspectives in order to damage their credibility.**

3. **Denies disfavored publishers from realizing advertising revenue they contracted for.**

4. **Demotes its competitors in online searches, so Google's offerings appear higher in listings.**

Google's power is bound to grow mightier still. With Google's vast data trove feeding ever more sophisticated artificial-intelligence algorithms, the search giant's ability to squash any new competitor will strengthen. When it can't beat competitors, it will buy them to silence them, as it has done more than 200 times since going public.[110]

As in business, so in politics, as they say.

And so, we add that a team of independent and apolitical researchers at the University of Chicago's Booth School of Business studied the impact Google has had on elections worldwide. It turns out that Google is not only influencing US elections to an alarming degree. It has become a global influencer. In random queries conducted with 10,000 people in thirty-nine countries, the researchers found: "[Google] has likely been determining the outcomes of upwards of 25% of the national elections in the world for several years now, with increasing impact each year as Internet penetration has grown."[111]

Twitter Burying the Tweets of Political Opponents

Taking Sides in the 2016 Election

Since its launch in 2006, Twitter has evolved from a quirky social hangout for young people into the world's "public square" where you'll find breaking news from the boring to the bombshell. The interesting thing about Twitter is how everything is trackable.

Each 140- or 280-character post, along with likes and retweets, can be seen and analyzed easily. For everyone from pool cleaners to politicians, it's a great tool for testing the marketplace, inexpensively learning whether the dog will eat the dog food, as the marketing people say. But if you are a conservative, you may be in for a shock—the dog can't find the bowl.

At the height of the 2016 election season, when millions were turning to this new public square to find the latest news on the two campaigns, some of that news simply was not appearing. Twitter was burying tweets—intentionally.

You'll recall the campaign. It was raging hot and bitter. Like a fever gripping the country and ripping at it, as well, with all sanity appearing to fly out the window.

Suddenly Hillary Clinton's campaign manager, John Podesta, finds his emails hacked. Those emails are released publicly—though they had been very private and very damaging to Clinton.

Too damaging in Twitter's view—so they started twisting the wheel to righty-tighty, and within minutes a firehose of tweets about the emails had been crimped to a garden hose dripping…

We surely would not have known anything about this, except that Twitter's former CEO Dick Costolo thought he could trust his employees in a pep talk he was giving one day. But one of the employees leaked a rather revealing excerpt of Costolo's talk:

> We suck at dealing with abuse and trolls on the platform and we've sucked at it for years. It's no secret and the rest of the world talks about it every day…We're going to start kicking these people off right and left and making sure that when they issue their ridiculous attacks, nobody hears them.[112]

After that bombshell became public, it was just a matter of time before Twitter's acting general counsel Sean J. Edgett would be giving written testimony to the Senate Committee on the Judiciary. On October 2017, he wrote:

Approximately one quarter (25%) of [#PodestaEmails tweets] received internal tags from our automation detection systems that hid them from searches…Our systems detected and hid just under half (48%) of the Tweets relating to variants of another notable hashtag, #DNCLeak, which concerned the disclosure of leaked emails from the Democratic National Committee.[113]

So Twitter did hide a quarter to half of all tweets that might have damaged Clinton even more. Edgett owned up in what we can presume was a kind of apology: We hid the tweets "as part of our general efforts at the time to fight automation and spam on our platform across all areas."

While most people had no idea this kind of thing could happen, it then begs the question: Did Twitter's engineers also suppress up to half of the automated spam-load of tweets that were praising Clinton for other, more laudable aspects of her career, such as her accomplishments as secretary of state?

Turns out, no.

Banning Is One Thing, Shadow Banning Quite Another

It's known as "shadow banning" and it's a kind of ban that's not apparent to the user. The user can go on posting like normal, but their posts cannot be seen by anyone but themselves. The profiles continue to appear when conducting a full search, but not in the autopopulating dropdown bar. That's what makes it so insidious. Unsuspecting searchers don't see the results where expected in the dropdown bar, and give up on the search.

It results in less engagement, fewer followers and retweets, but the shadow-banned individual doesn't know why.

Officially, Twitter insists they *do not* shadow ban:

"You are always able to see the tweets from accounts you follow (although you may have to do more work to find them, like go directly to their profile). And we certainly don't shadow ban based on political viewpoints or ideology."[114]

Of course they say that. What are they going to say, "We use our platform to advance our own narrow interests?" Don't think so.

But several Twitter employees were not so guarded.

We learned as much from James O'Keefe and his Project Veritas. In the same commando style that has elevated his political performance art to the heights the Left had long owned, O'Keefe went undercover at a San Francisco bar with some Twitter employees, for whom beer is a rather potent truth serum, it appears.[115]

Here are some of the more damning excerpts from O'Keefe's outstanding report:

Clay Haynes, senior network engineer:

"What we can do on our side is actually very terrifying."

"I'm pretty sure every single employee at Twitter hates Trump."

Mo Norai, former Twitter content review agent:

"If it was a pro-Trump thing and I'm anti-Trump…I banned his whole account…it's at your discretion."

When asked if the banning process is written policy or unspoken policy:

"A lot of unwritten rules, and being that we're in San Francisco, we're in California, very liberal, a very blue state. You had to be…I mean as a company you can't really say it because it would make you look bad, but behind closed doors are lots of rules."

When asked if left-leaning content was less scrutinized:

"It would come through checked and then I would be like, 'Oh, you know what? This is okay. Let it go.'"

When asked what impact this shadow banning had on the election:

"Twitter was probably about 90 percent anti-Trump, maybe 99 percent anti-Trump."

Olinda Hassan, responsible for Twitter's rules and regulations:

"We're trying to 'down rank' [so] shitty people do not show up."

Pranay Singh, Twitter direct messaging engineer:

When asked if control over certain political viewpoints can be automated via machine learning:

"Yeah, you look for Trump, or America, and you have like five thousand keywords to describe a redneck. Then you look and parse all the messages, all the pictures, and then you look for stuff that matches that stuff."

When asked who the algorithms are targeted against:

"I would say the majority of it are for Republicans."

"Just go to a random [Trump] tweet and just look at the followers. They'll all be like God, 'Merica, like, and with the American flag, and, like, the cross. Who says that? Who talks like that?

"There are hundreds of thousands of them, so you've got to write algorithms that do it [the censoring] for you."

Abhinav Vadrevu, former Twitter software engineer:

"It's risky, though. Because people will figure that shit out, and be like… you know, it's a lot of bad press if, like, people figure out like, you're shadow banning them. It's like, unethical in some way."

Set aside the level of real intelligence their recorded conservations suggest— they'd been drinking, they were partying among their own, or so they thought. If they are to be believed—and there's not much upside for them in lying since it puts their job on the line—Twitter not only bans users it doesn't like, it shadow bans them—and has been doing it clandestinely for years.

Going forward, it appears that either we are going to live with a very biased public square, or we have to find a way to reconstitute Twitter.

O'Keefe believes that as a minimum first step, Twitter must become more transparent:

Anonymity of its internal policies have bred irresponsibility and abuse. If Twitter wants to convince its users it truly respects free speech, there must be some transparency. Bring shadow banning out of the shadows. Algorithms are only as good as the weights put on them. So, take your thumb off the algorithmic scale. Get rid of the engineers who abuse their power and show us your HR policies.

O'Keefe's one-line conclusion should strike a chord with every American, regardless of political leanings:

> **"What kind of world do we live in where computer engineers are the gatekeepers of the 'way people talk?'"** [116]

How Many Conservatives Did Twitter Shadow Ban?

Apparently eighty-two pages worth. This damning indictment came from a document Twitter bosses never wanted anyone to see. It's clear evidence that Twitter tried to turn the 2016 election by censoring the posts of politically inconvenient users. These eighty-two pages were hacked and turned over to *Breitbart News*, and confirmed by another major publisher, the story said. [117]

A key to the excerpted page was provided by *YourNewsWire*:

> For the time being, only a partial list of the most notorious accounts is being provided due to the literally thousands of accounts on the list. The list appears to have been created more by an algorithm using filters for "buzzwords" than just popular accounts (this assessment was made due to the fact that an entire portion of the list contained back-to-back "deplorable" accounts with either deplorable in the user name, or profile description). Twitter also uses a filter that has an "admin delay" in which a post is delayed for view until its filter approves the post. If the person is low priority, it will normally take only 15-20 seconds once it runs through the buzzwords. If it's high profile, it may take manual approval. The "impressions" are reduced with a limit capacity which is why you're seeing less and less likes and retweets from conservative accounts: because limiting impressions means less people seeing tweets from these accounts in their regular news feeds.

In an effort to validate this eighty-two-page blacklist, *YourNewsWire* Editor-in-Chief Sean Adl-Tabatabai ran down some names on the list. One that he found was Mike Cernovich *(@Cernovich)*, the fourteenth name from the top of the list and owner of the website Danger & Play. Cernovich produced screenshots of his analytics tools, showing that he gained an average of five hundred followers a day until March 11, 2017, when those numbers showed a dramatic decrease to about a hundred a day, with one of those days showing only twenty-eight new followers.

To go from a high of 919 to a low of twenty-eight just nine days later shouts "Big Problem!" Likewise, he saw a 30 percent to 50 percent decrease in impressions, which meant his retweets were not being seen by his followers.

Those impression numbers dropped from an average of 800 to 173 in just a month's time.

> An interesting aside comes to mind: Hillary Clinton's incessant carping about Russian state interference in the election now appears to pale in comparison to all the conservatives' voices being dinged on Twitter. Indeed, Donald Trump may well have won by much more, even capturing the popular vote—had Silicon Valley not been aggressively interfering in the 2016 election.

Taking Aim at Prominent Republican Party Leaders, Too

Pick up any left-wing journal and they'll insist that claims by Republicans of shadow banning are "fantasy conspiracies" and "totally fabricated." Reach a spokesman at Twitter and they have a prepared response that is at least more nuanced. Along the lines of: "When our algorithms detect hateful, bigoted, or otherwise inappropriate behavior in an account, our algorithms downgrade the visibility of those accounts."

But it appears neither the knee-jerk explanation nor the nuanced one are any more than thinly spread cover for a political agenda. There's just too much proof.

We could understand Twitter trying to clean up the swamp they created. But when you suppress people like Republican National Committee Chairwoman Ronna Romney McDaniel, the explanations fall short.

Searching for McDaniel's account on Twitter was next to impossible for a while. Write in her name, and fake accounts would show up. But her real account wouldn't. Even though 80,000 people follow her on her verified-legit Twitter page.

Along with McDaniel, these members of Congress—Mark Meadows, Jim Jordan, Devin Nunes, and Matt Gaetz—have been shadow banned. Also feeling the sting is Donald Trump Jr.'s spokesman Andrew Surabian. All high-level public officials, obviously, not the kinds of trolls that infest Twitter.[118]

When it happened, President Trump took to Twitter to denounce shadow banning, vowing to "look into this discriminatory and illegal practice at once!"[119]

Gaetz responded by filing a complaint against Twitter with the Federal Election Commission. He cites, as grounds for his complaint, that Twitter's shadow banning "gives his political rivals an unfair advantage" which amounts to an in-kind campaign contribution to his rivals.[120]

Gaetz's office believes the banning was a negative retaliation for the hard-hitting questions the lawmaker put to Twitter executives in hearings. Gaetz made the case, soon to become a defensible legal case, that if social platforms such as Twitter want to continue enjoying the benefits granted to them by the 1996 Communications Decency Act, they cannot pick and choose which voices get heard.

"We do it because we hate conservatives (and because we can)."[121]

As we've seen with Facebook, Twitter is also two-faced when it comes to deciding who will be allowed to advertise on their platform. Twitter allowed Planned Parenthood to run ads promoting the right to abortion, but when Live Action tried to run ads promoting the belief that a baby in the womb deserves life, they were denied.[122]

Twitter's explanation? The pro-life ads are "offensive and inflammatory."

Live Action founder Lila Rose says her group has been "totally banned" from advertising on Twitter for three years. This despite having close to 70,000 followers and a Facebook page with two million followers. At the same time her group was being shadow banned, pro-abortion Planned Parenthood was permitted to advertise on Twitter.[123]

It's obviously an arbitrary decision the social platform is making—informed by a single worldview. But the arbitrariness of it is secondary to a larger theme. This is not about which side of the fence you are on. It's about the right that both sides have to freedom of speech in the new public square.

Live Action isn't the only group forbidden from talking about the right to life on Twitter. Marjorie Dannenfelser, president of the Susan B. Anthony List, also got slugged with the silicon fist. She was accused of violating Twitter's "hate, sensitive topics, and violence" policy.

And what post did she get flagged for?

A picture of Mother Teresa, much like the image created by the March for Life organization, with the caption, "Abortion is profoundly anti-women."

So the world's most celebrated Catholic nun and her kind words are just too…what?

What in the above image could so upset Twitter? It's a rhetorical question, of course. There's nothing upsetting about it, unless you support the mission of Planned Parenthood so fervently that you're all too happy to righty-tighty anyone who disagrees with Planned Parenthood.

We wonder. If the tables were turned, and Twitter refused to allow Planned Parenthood to run ads, would the streets fill with angry partisans shouting and sermonizing about free speech violations? Probably, as they rightfully should. Of course, it won't happen. Leftist partisans and Twitter have mirrored agendas.

It's as if they are saying, straight-faced, "We are trying to tamp down on hate speech, and since everything that conservatives say is hateful, they must be banned, right?" And if you are not one of them and you just stand there, waiting for them to burst out in laughter, slapping each other on the back for such a good joke—you might have a long wait. They believe what they say.

And they hold the keys to the new agora. It's as if we've returned to a bygone public square—the company towns, privately owned, privately policed with brutal frequency. It was an unhealthy way to run a republic—which is why our nation put an end to those company towns once before.

Enforcing Censorship—For Tech Tyrants, It Begins At Home

It was a hot story of summer, 2017. The story of Google firing white men for suggesting in internal emails that diversity programs ought to include diverse viewpoints. That is, the diversity programs ought not to only advance the leftist views of Google management. And with that kind of suggestion, all hell broke loose.

First, a quick backstory: A Google engineer named James Damore had fired off an internal memo calling for an open discussion of a number of things—among them the company's diversity policies, genetic differences between men and women, and a stilted working environment. Damore's principal concern was about what he would later call "Google's ideological echo chamber" that required conformity and diversity training along with internal seminars run by performance artists attacking white male privilege.

Damore was either the most honest fellow, or the stupidest person in Silicon Valley, career-wise. He must have known the discussion he sought would never take place, and the prospects of him continuing as an employee were also "never." Unsurprisingly, he was soon fired.

The reason for Damore's dismissal, according to Google, was his insistence on "perpetuating gender stereotypes." It was an odd rationale and even ironic, given that Damore's ten-page memo was complete with "suggestions" for improving the Google workplace. Suggestions such as this one:

> We have an intolerance for ideas and evidence that don't fit a certain ideology. I'm also not saying that we should restrict people to certain gender roles; I'm advocating for quite the opposite: treat people as individuals, not as just another member of their group (tribalism).[124]

That's the kind of talk that sent Google employees ballistic! Why, the very idea of "treating people as individuals" was so offensive to Damore's fellow employees that they demanded his immediate termination, along with a purging of anyone sharing his views.

And so next on the chopping block was David Gudeman. He'd previously got into a heated conversation with a Muslim co-worker who maintained he was being profiled by Homeland Security because of his religion. But in Google's eyes, Gudeman was *actually* accusing the Muslim of being a terrorist, and so had to go. Instead of dealing responsibly with two people's ideas, or even taking a side in the argument, Google went nuclear and fired Gudeman.

Federal law doesn't prohibit employers from discriminating against an employee based on that employee's political or cultural views. Title VII of the Civil Rights Act of 1964 forbids discrimination in employment only on the basis of race, sex, national origin, and religion. So, finding themselves clearly wronged, Damore and Gudeman took their case to the California courts.

California Labor Code Section 1101 states:

> No employer shall make, adopt, or enforce any rule, regulation, or policy: (a) Forbidding or preventing employees from engaging or participating in politics or from becoming candidates for public office. (b) Controlling or directing, or tending to control or direct the political activities or affiliations of employees.[125]

Which is exactly what Google did.

In October 2018, the case ended up in arbitration. Of course, Google probably argued that their corporate workplace is respectful—free of harassment, intimidation, or unlawful discrimination. Or some legalese to that effect. Indeed, they will claim that they fired Damore and Gudeman strictly in order to enforce these policies. And yet...

The Base Contemptuousness of Google's Employees Surfaces

And yet entered into the court's records were copies of emails senior Google managers sent out companywide describing Damore as "repulsive and intellectually dishonest" and advocating physical violence against him. One lashed out, calling him "a misogynist and a terrible person" and threatened, "I will keep hounding you until one of us is fired. Fuck you." This is when, by the way, Damore turned his emails over to the folks in Human Resources.

He was fired shortly thereafter.

In all, a hundred pages of employee postings on Google's chat system were turned over to the courts for discovery. A full one hundred pages of foul trash talk and threats toward white male conservatives. Some of the doozies included:[126]

- "[Damore is] a cancer within our culture."
- "If you want to increase diversity at Google, fire all the bigoted white men."
- "By being a white male, you are in a privileged class that is actively harmful to others."
- "'America First' is a slogan for American Nazis…and you should absolutely punch Nazis."
- "[Everyone should] boo white-male hires."
- "[We should discuss a] moratorium on hiring white cis heterosexual abled men who aren't abuse survivors."
- "[We should] advertise a workshop on healing from toxic whiteness."
- "[We should] award 'peer bonuses' for speech attacking conservatives."

Adding to these embarrassing and quite scandalous admissions was an interview with a former Google employee on what he'd witnessed:[127]

- "After the 2016 election, we had an entire 'T.G.I.F.' dedicated to the election result, in which several of our top management gave emotional speeches as though the world was going to end and seemed to be on the verge of tears. It was embarrassing."
- "The worst part isn't the 'diversity.' It's the 'inclusion,' the banner under which they justify dangerous pseudosciences like unconscious bias and microaggressions, and try to make them company policy."

- "Google is run like a religious cult. Conform and carry out the rituals, and you'll be rewarded and praised; ask any uncomfortable questions or offend the wrong people, and the threats and public shaming will be swift and ruthless. The religion in this case is a kind of intersectional feminism, its central tenets are Diversity and Inclusion, its demonic enemy is Bias, and its purifying rituals include humiliating forms of 'training' that resemble Maoist struggle sessions. This might sound crazy to a lot of...readers, but college students should understand, since it's a similar culture."

- "The agitation ranges from very subtle ('it's not OK...we cannot stand for this...these are shitty opinions') to quite overt ('this is violently offensive...I will not tolerate...I could not in good conscience assign anyone to work with you.')"

- "I've seen around twenty to thirty managers agitating this way, each of whom is in charge of anywhere from a few dozen to over a thousand employees. There are some very high-level people who consider the progressive agenda to be more important than the success and mental health of their teams."

Google managers also kept blacklists of conservatives and blocked them from working on their teams—effectively deep-sixing their careers. In one particularly revealing memo, a manager wrote about an intern who wouldn't soften his politics: "I don't think he was aware that there were real consequences for his actions. That might have given him the right motivation to change his beliefs or at least to keep his mouth shut."[128]

Once again...

Intolerant, Yes—Also Deciding What Most Everyone Sees Online

Another Google engineer who joined the class-action lawsuit is Manuel Amador, a systems engineer who was anonymously accused by a fellow employee of saying that intelligence and race are linked. This charge came out of left field for Amador, a Hispanic, because he had never said or written any such thing. What's more, he didn't believe in its truthfulness.

Turns out, the accusation was entirely fabricated in order to harass Amador. Despite that, Google sided with the harassers and asked that Amador issue an apology for something he had never said. Knowing what he was up against—that Google would go on allowing and even abetting this kind of hostile, tyrannical behavior from its employees—Amador quit the company instead. Following his departure, he released an open letter with some interesting insights:

> Google employs a few individuals (from rank-and-file to upper management) who are or have become highly ideological. They have made it one of their ostensible missions to have the entire company conform to these ideologies. Most of them believe that all of us—me and many others included—should not be permitted to impugn or question the ideologies they want to impose…
>
> Many people (including me) have faced contempt, opprobrium, insults, smears, provocations, threats of industry blacklisting, and even frivolous H.R. reports that influence my career (and many others'), in retaliation for voicing my mind. The tone of this treatment was always particularly intense whenever I dared to question the set of ideologies that I found incorrect, toxic or divisive. I have been slurred as a racist, a sexist and "privileged," in direct contradiction to the content of my thoughts…I have been directly ordered by senior management to "stop posting immediately."[129]

Yes, this is what's happening in the company that controls the information flow.

These are the actions of people who now dictate what up to 90 percent of Americans see online.

A former Google employee, speaking on the condition of anonymity to *Breitbart News,* explained how Google gets away with purging "wrong-thinkers":

> They use a firing process that enables them without exposing their real— political—reasons for doing so. Once the company has decided they don't want someone around, the usual way to get rid of them is to put them on a PIP ("performance improvement plan"), which highlights areas where they aren't living up to the formal job expectations for their position and sets specific goals for them to achieve. The expectations are written vaguely enough in the first place that a manager can always come up with an argument that someone isn't quite meeting them. If those expectations were enforced to the letter, almost everyone would be fired after their first review. The whole process takes 2-3 months, and it's basically just a way to generate a paper trail to guard against wrongful termination suits. If

someone claims they were fired for discriminatory reasons, Google can point to the PIP as evidence that they weren't.[130]

But in the case of Damore, Google slipped up, the informant believes. Hopefully, Damore's legal team was able to successfully represent him in arbitration with Google.

Not Solely Discriminating Against Conservatives

Beyond the ideological echo chamber of Google and much of Silicon Valley, there is also a deeply misogynist "brotopia" culture as chronicled in *Wired's* "'SEX PARTY' OR 'NERDS ON A COUCH'? A NIGHT IN SILICON VALLEY"[131] and *Vanity Fair's* "'OH MY GOD, THIS IS SO F---ED UP': INSIDE SILICON VALLEY'S SECRETIVE, ORGIASTIC DARK SIDE."[132]

Which is surely why, within weeks of the men suing, former Google female employees felt empowered to sue as well—for sexual discrimination. They, too, should have a case. Especially given the precedent of the US Labor Department's 2015 lawsuit against Google. The Feds analyzed the data on 21,000 Google employees and found Google guilty of "systemic compensation disparities against women pretty much across the entire workforce."[133]

In fact, most employees and executives in Silicon Valley are white, Indian or Asian men. And you can spend hours on TED Talks listening to these liberal-minded techies giving inspired TED Talks about saving humanity and the whales too…all the while treating their female employees like lower-caste "invisibles."

The Ignored Flood of Big Tech Censorship

Over the long history of social media history (over a decade, whoa!) we've seen the leading minds in technology devolve from a beautiful libertarian worldview—"let a billion ideas bloom on our platforms, it will be splendid"—to a bitter authoritarian view—"we know what should be said on our platforms, it will be so decreed."

How it happened, how a couple thousand computer nerds in Silicon Valley went from this broad-minded embrace of technology's potential to a near intolerance for any views other than their own—well, that will surely be the study of sociologists for decades to come. As for now, these engineers are the ones orchestrating the social tastes, political preferences, and informational choices of America and much of the world.

"One man's fake news is another man's gospel truth."
—DAVID WEBB, CEO OF TIMERA MEDIA. AND HOW IS GOOGLE TO DECIDE THAT? [134]

Where Does This Path From Libertarian to Authoritarian Lead?

This transition from a libertarian to an authoritarian mindset has not happened overnight. It has been steadily escalating. And so, like the residents of homes built on flood plains, assured by shyster realtors and tax-greedy city councils that growth is good, and storms rarely happen, and believing that emergency services will be there to rescue them should the waters rise to their rooftops, we haven't thought to prepare ourselves and leave for higher ground when the lightning flashes and the thunder rolls. But the flood will inevitably come—the waters will inevitably rise, the residents will ultimately drown under the waves of uncontrolled and deceitful authoritarianism. One by one, like those residents fooled into staying put even as the waters rise, we will see dissenting views threatened by the Tech Tyrants. One by one, these views will be suppressed.

If the Big Tech Tyrants are allowed to continue hiding behind the antiquated protections of last century's communications regulations, the rising waters of suppression and censorship will continue, and no doubt escalate.

Who, or rather what, could the Tech Tyrants elect to silence next on their social platforms?

Only one view of vaccinations, for instance?

Good and thoughtful people can—and emotionally do—disagree about the benefits and potential dangers of today's vaccines. This is because medicine, a branch of science, is always learning, always advancing. And it's important that all points of view find purchase in the online public square, so these ideas can clash and advance our understanding of this vital health issue. Yet the Social Tyrants have rallied around the "all vaccines are good" thinking. So, will any opposing views begin disappearing from the platforms?

Only one view of home/charter schooling?

As they have on the issues of immigration, deficit spending, relations with Russia, and free speech itself, the two political parties have switched sides in the debate over home schooling and charter schools in recent years. Angry partisans in both parties are for the most part oblivious to the history of the issues they now argue with every *ad hominem* tool handed to them by party leaders. But at least they can argue.

Young parents have strong opinions on the use of taxpayer money for public schools vs. charter or home schooling. Few in Silicon Valley and Seattle (Amazon and Microsoft) have steadfast opinions on this, because their children are enrolled uniformly in the nation's most prestigious prep academies.

But the Tech Tyrants can be counted on to toe the *current* Democratic Party line. Which means whatever the teachers union says, it should be. Which means charter/home schooling must be picketed, protested, shouted down, and discredited by whatever means possible, including insisting loudly that any parents seeking a higher-quality education for their children are misguided in leaving the public school system. Is this going to be the only view a young parent will find in the online discussion groups?

Only one view of CBD and THC medicine?

With cannabidiol oil and cannabis medicine becoming popular, few know that they used to be extremely popular. In the 1800s, they were the active ingredients in thousands of medicines. The same for other natural cures and herbal remedies. But the discovery of aspirin followed by the pharmaceutical industry's sudden rise into existence effectively buried these "potions" of old. Will it happen again? Should it? Good questions to debate in the public square—absolutely. Not questions to be decided by the whims and fiat of the online platform operators.

On and on we could go through the list of always-contentious issues—our Bill of Rights protections, guarding our national borders, the value of the Second Amendment, definitions of life and the right to life—all issues that could be swept in one direction by the social platform operators orchestrating the big flood.

The one we now see coming; but will we react in time?

The Stealth Takeover by the "Four Comma Club"

When, in August 2018, Apple became the first company in history valued at more than a trillion dollars—that's twelve zeros and four commas—they popped the corks in Cupertino, and investors who've held Apple stock since its

1980 IPO were woozy, with every $1,000 they invested now worth over a half million dollars.[135]

A month later, when Amazon followed, there was more cork popping and even happier investors who could have seen every $1,000 investment turn into $1.3 million just two decades later.[136]

Apple and Amazon are the world's first "four comma" companies in investors' eyes because they are the strongest of the monopolies, with the most devoted fan base, so they can get away with a lot more than the other Tech Tyrants.

But they are technically not the first into the Four Comma Club.

It's more accurate to say that this club—which has a roving clubhouse but no official charter or rules, and has about 1,500 members worldwide—has become the world's de facto governing body. Most of these 1,500 members, all multibillionaires on paper, have fortunes tied to Silicon Valley and Seattle. So, while all but Apple and Amazon are merely three commas deep on their own balance sheets, they are in a position to effectively control 97 percent of the world's wealth.

Not Your Daddy's Robber Barons

The Four Comma Club's power is being exercised almost capriciously—free of the old constraints of the capitalist world because those constraints are no longer being enforced in any meaningful way.

We could fill another book with stories of the Four Comma Club run amok. We'll limit it here to choice examples of Tech Tyrants actively undermining our democracy because it suits them—like little Lord Farquaads pissing over their castle walls on all below.

In Farquaad's time—and by that we mean to invoke a feudal past we were supposed to have advanced beyond—the kings in their kingdoms would hole up behind castle walls and extract rent from the villagers nearby in exchange for some occasional protection and shelter whenever unfriendlies came marauding. The "castle" concept was a big advance over people's previous living conditions, actually. But it remained a brutal time for those stuck being villagers.

Is the approach taken by today's Tech Tyrants any different in its way?

Their big modern advance is the "platform." Their insight has been to turn everything that can be a platform, into a platform. As this infographic from *Visual Capitalist* shows, the Big Tech Tyrants have made short shrift of their daddies' rules of commerce.[137] In just fifteen years the household names of GE, Microsoft, Exxon, Citibank, and Walmart have all—except for the one tech firm on the list—been shriveled down to shells of their former captain-of-industry selves, replaced at the top by Apple, Alphabet, Amazon, Facebook, and Microsoft.

At the current turn rate, tech companies should succeed in platforming nearly every American industry near to the end of Trump's second term. Academics have estimated that this will result in the loss of seventy-three million jobs. Obsolesced, as they say.

Some would say this is the way of progress. But is it advancing society for the better?

When we look back at America's big public companies of the 20th century, we see that they employed millions in well-paying jobs, complete with opportunities for advancement and fulfillment. They were far from sainted corporate citizens, but they raised the standards of living of us all. That is *not* what the Big Tech Tyrants do.

They manufacture their products in overseas sweatshops and slave mining operations, rely on algorithms and automation to "efficientize" operations and downsize numbers of pesky human workers and concentrate wealth at obscene levels with the help of Wall Street quants and Washington wonks—all another day in the long con they're running on America.

There was a time when booms in industry, and even in the tech sector, would lift all boats. That's not how it works anymore. Now, only the boats of the richest rise with the tide, while so many other boats, anchored to the bottom by chains of debt, are inundated and left to sink. That's the underlying reason why people on both the left and right of the political spectrum are upset. That's in part why we're seeing a "techlash," why we're seeing people on the streets hurling stones at buses transporting Google employees to work from their homes in San Francisco.[138]

The stones are not being thrown solely by the disaffected and unfortunate.

They are being thrown metaphorically by economists. Tasked with explaining the growing divide between rich and poor in America, these economists are increasingly concluding that a few massive companies are using their size to crowd out all innovators—so the net effect, at least to the casual observer, is that these massive companies alone are the most productive and innovative when they have, in fact, done it by crowding out, silencing or scooping up all others. Some boats rise, many boats sink.

And this is a new phenomenon.

Going back to the industrial revolution, with each new breakthrough or innovation, the benefits diffused out across companies and industries, with everyone benefitting to a degree. That is no longer happening. Today's engine of growth is fueled by technology's advances, but it is not translating into productivity gains for anyone except the top tech companies.

These productivity gains—such as more output per hour per worker—are the very thing that drove our collective living standard higher for centuries. But economists are finding that today only the top companies are getting more productive. Everyone else is in trouble and the gap between the two is widening.

As Andrew Haldane, chief economist at the Bank of England, told reporters at *The Wall Street Journal:* "Whatever good stuff is happening at the high end is not diffusing down to the tail."[139]

Since the 2008 financial crisis, overall US productivity has grown by about 1.2 percent a year. That's half the rate of the 1970s and only one-third of the post-World War II period, once adjusted.[140] Some have blamed this productivity slowdown on:

1. **Super-low interest rates;**

2. **Not knowing how to measure the output of a newly digital world; and**

3. **A long-term secular decline in our innovativeness.**

All three of these play into the productivity issue, yes. But none explains why "the gap" between the top performers and all others is so wide, or why the overall economic strength of the nation is in steady decline.

For that explanation, look to the top companies which have automated their operations on a global scale, eliminating almost all employee costs, squashing any emergent innovation before it takes root. Some have called this a winner takes all game. Others have called it monopolists run amok. One thing is certain:

The Big Tech Tyrants and their platforms are the new castles and only a few are lucky enough to live in them, with most stuck outside in steadily declining conditions.

Telling the US Government To "Pack Sand"

It has become the fashion for employees at Google, Microsoft, and Salesforce—because they see themselves as global citizens beholden to the innovative interest only, certainly not to the American interest—to refuse to work on their employers' government contracts.

Google's employees went into open revolt over the company's decision to provide AI to the Pentagon. Working for the government, they say, "is an ethical litmus test."[141]

Both Microsoft's and Salesforce's employees, hundreds of them, formally objected when their companies obtained contracts to help US Immigration and Customs Enforcement (ICE). They wrote letters to their CEOs condemning the Trump Administration and insisting that they understood better how this country should be run. An excerpt from the Microsoft letter: "We believe that Microsoft must take an ethical stand, and put children and families above profits...As the people who build the technologies that Microsoft profits from, we refuse to be complicit."[142]

When not just a few but many thousands of individuals within major companies—however poorly informed these individuals are—put their own beliefs before the company's mission and the nation's mission, there is reason to question whether the products made by those companies might be compromised.

And reason, as well, for a potential investigation for corporate malfeasance.

Openly Publishing Lies About Political Opponents

In the 2016 presidential election, 95 percent of Silicon Valley employees boasted of their loathing of the Republican nominee as measured by their financial contributions.[143] Then, even while Donald Trump was accepting the nomination for President, 145 senior tech executives launched a very public and scathing attack on Trump. It was an unprecedented action—both in the sweep of its contemptuousness, and in its essential fatuousness.[144]

It is worth revisiting the open letter these Big Tech Tyrants ran on July 14, 2016—and doing a little fact-checking now that President Trump's first term is half over.

We are inventors, entrepreneurs, engineers, investors, researchers, and business leaders working in the technology sector. We are proud that

American innovation is the envy of the world, a source of widely-shared prosperity, and a hallmark of our global leadership.

TRUTH: Silicon Valley is located in the US, but it is technically stateless and in the minds of its inhabitants, a ship unto itself.

In fact, American innovation often lags the rest of the world and is a source of prosperity to only a lucky few.

We believe in an inclusive country that fosters opportunity, creativity and a level playing field. Donald Trump does not. He campaigns on anger, bigotry, fear of new ideas and new people, and a fundamental belief that America is weak and in decline. We have listened to Donald Trump over the past year and we have concluded: Trump would be a disaster for innovation. His vision stands against the open exchange of ideas, free movement of people, and productive engagement with the outside world that is critical to our economy—and that provide the foundation for innovation and growth.

TRUTH: Silicon Valley likes open borders and lopsided trade policies because they can make a lot more money as a result. And safe behind their own security systems, they never have to deal with the consequences of open borders, or care about those who do.

Let's start with the human talent that drives innovation forward. We believe that America's diversity is our strength.

TRUTH: Such a sweet thing to say. Yet even at "enlightened" outfits like Facebook, only 4 percent of US employees are Hispanic and only 2 percent black. Only one in three jobs is held by a woman—and their pay seriously lags behind men. In fact, the brazen misogyny of the Valley is its biggest open secret.[145]

It's one thing for the engineers of the Valley to be backward sexists and racists in their own lives, it's quite another when they build that misogyny into technology. Which is exactly what we are seeing them do.

Google's photo app, for example, is learning how to automatically label digital pictures. It classifies images of black people as gorillas. Is this a case of the AI learning, improving at a difficult task? Or is it more a result of the kinds of people who programmed the AI and their own true views on the world?

Either way, releasing such a product into the marketplace is an inexcusable mistake. The nonprofit ProPublica studied this and found:

Widely used software that assessed the risk of recidivism in criminals was twice as likely to mistakenly flag black defendants as being at a higher risk of committing future crimes. It was also twice as likely to incorrectly flag white defendants as low-risk.[146]

Many police departments across the country have already purchased this software and use it to determine which neighborhoods they should police. So real people's lives are affected by this. And while surely some of the engineers working to perfect these products are very much concerned about the rush to market, others are no doubt laughing in their cubicles about the impacts of their products in the real world—since it is not likely a place they will ever visit or care about. "It's just that 'Merica place, so who cares?"

Great ideas come from all parts of society, and we should champion that broad-based creative potential. We also believe that progressive immigration policies help us attract and retain some of the brightest minds on earth—scientists, entrepreneurs, and creators. In fact, 40% of Fortune 500 companies were founded by immigrants or their children. Donald Trump, meanwhile, traffics in ethnic and racial stereotypes, repeatedly insults women, and is openly hostile to immigration. He has promised a wall, mass deportations, and profiling. We also believe in the free and open exchange of ideas, including over the Internet, as a seed from which innovation springs.

TRUTH: Trump ran a coarse campaign—how else could he have broken through the two-party stranglehold on elections? His coarse behavior has continued into his presidency; it's his style, sort of like Lyndon Johnson's or Andrew Jackson's actually. But to the point, Donald Trump has yet to disallow immigration that's legal. That dubious distinction belongs to the politicians on Capitol Hill who devised a system whereby people who try to honorably immigrate will struggle for up to sixteen years to gain citizenship.

Donald Trump proposes "shutting down" parts of the Internet as a security strategy—demonstrating both poor judgment and ignorance about how technology works.

TRUTH: What then of Facebook's and Google's now-proven campaigns to "shut down" conservative voices on their platforms?

His penchant to censor extends to revoking press credentials and threatening to punish platforms that criticize him.

TRUTH: Everyone who has ever run for president, including Trump's predecessor, has been driven crazy by the exasperating strain of running the media gauntlet. No press credentials have been revoked, barring one: CNN's Jim Acosta stepped over the line with his combative and rude behavior, and the White House was justified in temporarily revoking his press credentials. Meanwhile, censorship is flourishing online—enforced by the social platform operators.

In closing their letter, the Tech Tyrants wrote:

> Finally, we believe that government plays an important role in the technology economy by investing in infrastructure, education and scientific research. Donald Trump articulates few policies beyond erratic and contradictory pronouncements. His reckless disregard for our legal and political institutions threatens to upend what attracts companies to start and scale in America. He risks distorting markets, reducing exports, and slowing job creation.
>
> We stand against Donald Trump's divisive candidacy and want a candidate who embraces the ideals that built America's technology industry: freedom of expression, openness to newcomers, equality of opportunity, public investments in research and infrastructure, and respect for the rule of law. We embrace an optimistic vision for a more inclusive country, where American innovation continues to fuel opportunity, prosperity and leadership.

TRUTH: Big Tech Tyrants want a candidate who will help them ride roughshod over the spirit and even the letter of existing antitrust laws so that they can continue taking over one industry after another—relieving millions of business owners and formerly well-paid employees of their wealth—while stuffing and lining their own pockets with it.

Now we are over two years into Donald Trump's presidency. And we haven't seen the horrific scenarios the Tech Tyrants and Four Comma Clubbers so dramatically predicted. But, of course, that has not changed their tune. They continue to believe the pair of twos they've been dealt will magically become a full house or, better, a royal flush. And it is Trump's fault, they perversely rationalize, for not dealing them a better hand.

The Big Tech Tyrants remain 95 percent opposed to whatever Mr. Trump does—as if stuck in a techno tar pit of their own making. And they will continue to erode—the foundations of our free democracy.

Cornering the Currency Market of the Future?

In early 2018, Facebook and Google abruptly halted every cryptocurrency company from being able to advertise on their platforms. Each platform's official position was: We are trying to protect our users from "financial products and services that are frequently associated with misleading or deceptive promotional practices."[147]

It's certainly true that the fledgling cryptocurrency industry is brimming with con artists and flim-flammers (like all startup industries always have). But if the social platforms were truly devoted to protecting users, why did they keep mum long after they learned about the bad-acting Russian moles running political ads in the 2016 election, saying nothing until forced to?

Was the real reason Facebook and Google shut down the cryptocurrency ads to do a Tonya Harding-style kneecapping of the upstarts in this new industry? As it turns out, Facebook had just formed a secret group to develop a blockchain product that included its own cryptocurrency. Plus, Google had approached Ethereum founder Vitalik Buterin, soliciting his help in developing a cryptocurrency project.[148]

What's more, Amazon and Microsoft are also developing blockchain and cryptocurrency products, and they too have banned cryptocurrency promotions. Put differently, four of the world's eight largest companies appear to have formed an online-advertising oligopoly that is openly denying potential competitors access to the platforms that control upwards of 90 percent of the online information flow. Essentially silencing them.

If the social platforms were banning cryptocurrency companies from advertising because they wanted to kibosh competition—that's illegal.

Sure, the social platforms will argue that they are protecting users, not thwarting competition. But they've long ago exhausted any grace in their rationalization sprees.

Indeed, in an opinion article in *The Wall Street Journal* seeking to shed light on these bans, antitrust expert Mark Epstein called them "dubious Bitcoin bans." And he cited a 1951 US Supreme Court decision, *Lorain Journal v. US, in which the courts said a* company has a right to "refuse to accept advertisements from whomever it pleases" but it cannot refuse ads in order to harm its competitors.[149]

In 1951 it was a Wisconsin newspaper on the hot seat; today it is social platforms. The two are not perfect analogs. But the outcomes of their actions are similar, and so the precedent should hold: The Tech Tyrants should not be able to use their platforms to silence rivals. Any company that is shut out by Google or Facebook will find it impossible to reach an audience of any scale. That's the plain fact.[150]

There have been more recent cases and Epstein cited them in the *Journal*. In 1990, the Supreme Court held in FTC v. Superior Court Trial Lawyers Association that "social justifications for the restraint of trade don't make the restraint any less unlawful." And the courts have used this ruling to nullify advertising bans that have been similar to the cryptocurrency bans.

So there is a case to be made against Google, Facebook, Microsoft, and Apple. A case made vitally important by the subject at issue—our nation's transition into a potentially amazing new technological era. As the eminent futurist George Gilder reminds us:

> This cryptographic revolution…endows individuals with control of their data, their identity, the truths that they want to assert, their transactions, their visions, their content and their security In charge of our own information people can be anonymous if they wish, while also letting them keep a time-stamped record of all their previous transactions. It allows us to establish unimpeachable facts on the internet.[151]

The Withering of Freedom and Liberty

"Freedom" and "liberty" have become as hackneyed and passé as the Miss America pageant in the eyes of today's cultural taste makers—the so-called elite, sophisticated society.

A quick search in Google's Ngram viewer, which catalogs the use of words in the millions of books that Google has scanned, shows how out of favor the word "liberty" has become over time.

Google's tool, while far from perfect, is a useful way to get a 30,000-foot view on things. And we can see the downtrend in interest and writing about "liberty" steadily dropping from a time only a few years after America first rang the Liberty Bell.

It is the truth of our times that many people hear the words "liberty" or "freedom" and get uncomfortable, or they cringe at hearing them, or think them passé along with the patriot baggage they presume.

In an odd way, this state of affairs can actually be viewed in a positive light—that is, as a tracking mechanism on the overall great state of the nation: In general, Americans feel safe and secure since they have their basic needs met. America's external enemies have become fewer, so folks have relaxed regarding the big life-and-death concerns and are focused more on cultural concerns. Sadly, many Americans—especially the young—are ignorant of history and human nature. They are unaware of how quickly their liberty can be stripped away by tyrants with dark minds and power-hungry hearts.

They've had the luxury of forgetting the battles previous generations fought over these two words: Liberty and Freedom.

They've been able to launch themselves into what they consider bitter cultural wars, left against right, etcetera, etcetera. They think these are tearing our nation apart. They think this because they have little real understanding of history and how these culture wars have always been rekindled in times of relative prosperity free of external threats. And they are likely to carry on in this thinking until our nation's economic engine seriously sputters, or we are weakened to the point that, historically speaking, an enemy will again present itself on our shores.

There is no reason to think it won't happen, or that our time will be any different.

And when it happens, then we think we'll see another side of the public mind. We'll suddenly remember what it means to flex a patriotic muscle and shoulder up with political rivals in defense of liberty and freedom.

Or many of us will, anyway. A majority of us. More than enough.

Until that time comes, we can each of us define these two words any way we chose.

At the core of these definitions and exercises of freedom and liberty is one central idea that has been with us since the beginning: basic privacy. Without it, everything falls apart. And with this basic privacy coming under such threat from the social platforms, what do we do?

It's time for a visit to the woodshed. Turn the page!

Out to the Woodshed

Business Ain't Beanbag or Baseball, but Three Strikes You're Out

We've seen what happens when a society lifts its technologists up onto a pedestal, elevating them to the stature of gods, and heaping constant praise on them for all the wonderful new gadgetry they give us.

We begin thinking these new gods—Google, Facebook, Amazon, Apple—are making our world a better place.

We marvel in their creations—brought forth *ex nihilo* as far as we mere neophytes are concerned.

We invest ourselves, our emotions, our families, our careers—our very society—and our money in each and every new miracle the gods deem worthy of our attention.

We admit openly of how much we love being the beneficiaries of so many technological advances—wondrous, productive, and entertaining—coalescing in our time.

But then, as reports trickle in, as rumors of deception fill the air, the veils of gold-laced and silicon-threaded curtains separating us from the gods are pulled aside, one by one, and we begin to learn that these platforms—these shiny totems—have been built on dark algorithms—obscure, sometimes unscrupulous tools of trickery and deceit optimized to administer steady injections of dopamine—like a chemical bullet to the brain—hooking us, addicting us in very real and often dangerous ways.

Which Is Strike One

Our eyes have been opened, and now we can clearly see how the social platforms were built on a foundation of well-planned privacy theft that is sadly little different from the practices of the old communist Stasi.

The social platforms' very business models require that they play fast and loose with the basic privacies upon which our nation was founded, privacies we cherish and hold to be most dear—each in our own way, which is itself one of the first freedoms.

By squandering our privacy, by surveilling and recording our every move, our every spoken word—revealing and contorting our very thoughts—the Big Tech Tyrants have not only dehumanized, manipulated, and monetized the citizens of the United States, they are plugging their machinery into the very power centers of the three branches of government and are attempting to move national decision-making out of the light of public domain, and install it behind the dark castle walls in Silicon Valley.

Which Is Strike Two

But we know the corporate world has always been ruthless; it's no place for loafers or lullabies. So we attempt to rationalize Silicon Valley's initial over-reaching excesses and promissory sleights-of-hand; we chide them for their boyish indiscretions, their sometimes churlish, often socially inept behavior, but we continue to applaud them for giving it their all. We tell ourselves we could go on treating them like mere platform performers and not the power-house publishers they have become. Perhaps, just maybe, if they are truly contrite, we could grant them immunity under the law for all the crazy sewage that spews from their platforms.

But we have now learned that they are not mere conduits for torrents of user-generated content.

All along they insisted their algorithms were impartial, their curation sincere; they were simply giving the people what the people wanted—in the best way they knew how. They were benevolent boy kings who had only the best of intentions.

But now we know better. Now our focus has cleared.

Now we understand the purpose of the veils.

For we are uncovering case after case where these unabashed, unrepentant, arrogant techno-plotters are actually forcing their own personal worldviews on the nation. All along, it turns out, they have been deciding what gets seen and what gets silenced.

Privately, surreptitiously, abetted by leagues of lobbyists, they have been—continue to be—dabbling in politics, turning their algorithms into vote-swaying machines.

Since they can control up to 90 percent of the information flow online, this means they can also have a nation-changing impact. They may think themselves high-minded in this—"all for the national good, you know"—but their actions are Techno-Stasi in a not-hyperbolic comparison.

Which Is Strike Three

If ever there was a time to take the Big Tech Tyrants out to the woodshed, that time is now. It is long overdue, in fact.

These Tyrants are still standing at the plate, leaning in, cocky, practically daring anyone to challenge their great power: their corrupting, corrosive power; their invasions of privacy; their control over the minds and tongues of the people.

They are growing richer and more authoritarian while all around them lies the smoldering detritus of millions of jobs and opportunities.

All because they can.

It took China 2,000 years to complete their wall around what would someday become a billion subjects, er, people. By comparison, it took Facebook less than a decade to reach a billion users, er, customers. Each was a fascinating achievement, but also dark...

How Different From the Tech Titans of Old

Walt Disney built an early tech company to provide all-new entertainment visions. As individuals ready to be thrilled by innovation in entertainment, we loved Disney. As a society, we loved Disney because they also employed 185,000 people—at one point more than ten times as many as Facebook.

Giants of the industrial age like General Motors, and tech companies like IBM, employed hundreds of thousands of workers. The spoils of these giants were carved up fairly. Investors and executives who bet on or ran these companies and their products and services made good money, yes, but their American employees could buy homes and motorboats and still afford college tuition for their kids.

Now, so many of those kids are looking at their lives through lenses of minimum wages or underemployment, or no job at all—for many of the job opportunities that once held doors open to America's young people have been closed in their faces, and the opportunities have been shipped offshore or automated into the ether by the Tech Tyrants.

Jobs and economic fairness, or their lack thereof, are at the core of the American problem today. There are more layers, of course; like an onion, you can peel away at issues such as globalism or immigration in a search to find scapegoats (or real enemies, foreign and domestic) to blame. And you can engage forever, and fruitlessly, in wondering about what went wrong with "the dream."

Why so many jobs have gone away.

Why the heartland economy is stagnant.

Why young people see their parents gazing wistfully out windows, shaking their heads, waiting for a bright future that was supposed to come…but never came for their kids.

And at the core of all this questioning you will find the Big Tech Tyrants. Now happily ensconced on the highest peaks of success, the Tech Tyrants have lost touch with the people in the valley of despair far below.

In all our questioning, all our handwringing, all our searching for some guilty party, we must look up to the summits where the Big Tech Tyrants enjoy their daily bounty. They are the ones who are primarily responsible for the turmoil raging down below. Is it not a little ironic that it is we who helped them build their Olympus-perched mansions?

These boy kings dumped enormous wealth into the hands of a tiny bunch of investors and talented engineers, leaving everyone else to cope with the dispersion and addiction of the new opiate of the masses they've created—a hypnotic stream of video content on super-fun phones.

They've become so powerful, so much more powerful than Hollywood, than Wall Street, than even Washington in their ability to dictate how we should all live our lives…that stock market investors have bid them up, up, up—counting on them to double or even triple in size in the next decade.

Power begets power.

And that unchecked concentration of power allows the Tyrants to further cement in place their own narrow political vision for America, and for the world.

As they also corner the market on the new technologies of machine learning, artificial intelligence, job automation, and blockchain, they are building economic barriers around themselves, new Great Walls—making them even more impervious to the outraged cries for reform shouted from down below by responsibly acting politicians or agitators like us.

However…

Every Castle in History Has Finally Been Breached

It's a fact (or what we used to call a fact before all meaning was siphoned from the term, thanks largely to the social platforms) that every assuredly impregnable castle in history was finally breached at some point. Whether Jericho besieged by Joshua, or Saddam's palace taken by the U.S. military in 2003 during the Iraq War, the walls all come tumbling down.

But how long before today's Big Tech Tyrants relinquish their hold over speech and privacy?

Will it be hundreds of years, like in times of old? Not likely, we think. So, maybe only fifty years, or ten years, since things move so fast now?

Meanwhile, those lodged in the stratospheric heights of today's silicon castles are laughing down on governments of, by, and for the people; they are crushing, or acquiring, any good business that might be competitive, trampling with impunity over us simple serfs and peasants with algorithms that smash through our social villages like armored knights on giant stallions.

It's like we've been tossed through some time portal into a nouveau-feudal period. And the castle walls, gleaming high above, keep getting fortified, growing taller.

Who can stop them?

> The past doesn't repeat itself, but it can rhyme. Standard Oil, AT&T, Microsoft—a century of Tech Titans dominating their markets so powerfully, action was required. Once again, we are there:
> - **Forty-three percent of Americans report getting their news online, and Google controls 65 percent of internet searches.**[1]
> - **Three out of every four dollars spent on digital ads in the US goes to Facebook or Google.**[2]
> - **Facebook products are used by 95 percent of young adults on the internet.**[3]
> - **Amazon accounts for 75 percent of electronic book sales.**[4]
> - **Google and Apple own 99 percent of mobile phone operating systems.**[5]
> - **Apple and Microsoft own 95 percent of desktop operating systems.**[6]

Forces Assembling for Change

In a time when the two political parties can't even agree on why they disagree, there is remarkable bipartisan support for overhauling the social platforms. Leaders in both political parties will never, of course, agree on how the platforms should be made accountable (from what we've seen during congressional hearings, very few of our elected leaders even know what a platform is, or what one does). They only know that some kind of accountability must cemented in place, and that the time to pour the concrete of accountability is now.

Left-wing groups

Leading the charge from one end is Freedom From Facebook, with their organizer, Sarah Miller. She made a splash at the July 2018 congressional hearings, holding up an image of Mark Zuckerberg and Sheryl Sandberg as a giant, two-headed octopus.[7]

There's rich American lore behind bankers and Republicans being subjected to the "giant vampire kraken" metaphor—but it was likely the first time leftist techies felt the comparison's sting. Especially since Freedom From Facebook was born out of the George Soros-funded Open Markets Institute and counts nine other leftist groups among its coalition, including the 700,000-strong Communications Workers of America union.

Nonpolitical insiders

From the vested middle—that is, engineers in Silicon Valley who are shamed for what they've wrought—there is a troika emerging from the talents and power brokers leading the call for reform.

Tristan Harris had been Google's design ethicist until he could no longer stand the deceptive practices Google used to literally addict people to its services. Harris spent hours in TV interviews detailing the "dark patterns" the social platforms used to hook people on smartphones, engineering those phones as if they were slot machines—the jackpots being "likes" rather than coins.

Roger McNamee had been a tech investor and one-time mentor to Mark Zuckerberg. He has said publicly that the Zuckerberg he knows will not only never fix Facebook, but that the young CEO truly doesn't understand why people are so upset about privacy.

James Steyer saw the problem developing back in 2012, and he laid it out in his book *Talking Back to Facebook*. His biggest worry was the impact the social platforms were having on children, and he began trying to bring about change in Washington. But for years he was counter-assaulted by the armada of lobbyists the social platforms deployed to drown him out in the halls of power.

Steyer's only real influence, however, was with the readers of his book. He was widely attacked and pilloried in Silicon Valley—though he was a Stanford professor. Even his book was almost censored. When he published it, his bene-factors at Common Sense Media—an educational kids' media company—were called onto the carpet at Facebook. There, Facebook COO Sheryl Sandberg and VP Elliot Schrage told them "we're big and important in Silicon Valley... maybe you don't want to be on the board of Common Sense if you want to be doing business in Silicon Valley." The book was going to cause problems, they said, and "they could cause problems too."

Schrage denied making any such threat, but there's no reason to doubt the truth of it.[8]

These are the most visible "white knights" leading the charge in Silicon Valley. There are more, of course, and the popularity of the #DeleteFacebook movement attests to the marketplace's efforts to rein in the Tech Tyrants.

Right-wing groups

As we've detailed in these pages, there is a big tent of support on the political right for an overhaul—because many conservatives, and especially conserva-tive publishers, have felt the brunt of leftist censorship on the social platforms, the reckless approach to privacy, and the indifference to the platforms' broader impact on society.

States Taking Their Own Action

With California in the lead, every state is looking at beefing up privacy protec-tion rules. Each state has its own focus—facial recognition uses, net neutrality, drone deployments, student learning, license plate readers, right to be forgot-ten, and more. With so many states acting, there is impetus for a single set of federal rules across all fifty states.[9]

European Community Not Fooling Around

In May 2018, the European Union slapped the largest fine (1.5 billion Euros) in history on Google and began mandating a whole new level of privacy. Social platforms operating on the continent would now be required to obtain users' permission before collecting their data. Consumers were given new rights to download or delete their information. Any violations would result in hefty fines.

At the core of this new General Data Protection Regulation, or GDPR, is a willingness to confront the economic and political challenge of our time:

How social platforms, like the oligopolies they are, have concentrated power and widened their profit gaps between the top few and everyone else, creating formidable barriers to entry, and chilling or killing innovation across industries.

Typically, Americans have idly laughed at Europe's economic ways, its statist views, and traditional slowpokeness—conveniently forgetting, as Americans are wont to do, that Europe has the Autobahn, high-speed trains, Ferraris, Mirage fighter jets, and the world's largest and fastest particle accelerator at CERN (not to mention nukes). Rana Foroohar, author of *Makers and Takers: The Rise of Finance and the Fall of American Business*, has poked some holes in these perceptions and shown that European markets are, in fact, more competitive:

> European institutions are more independent than their American counterparts, and they enforce pro-competition policies more strongly than any individual country ever did with "lower levels of concentration, lower excess profits and lower regulatory barriers to entry" and it's the huge rise in U.S. political lobbying as the key reason that levels of concentration between the two regions have diverged since the 1990s.[10]

And, with respect to Silicon Valley's constant snickering about Europeans not having an internet giant because they simply are not innovative, Foroohar adds: "U.S. tech groups have been using that tired old line in Brussels for years", apparently forgetting that it was a British computer scientist, Tim Berners-Lee, who invented the World Wide Web while working at CERN the European physics research laboratory.[11]

Democrats on Capitol Hill Wary of Biting Hand That Feeds

For their parts, most, though not all, Democrats on Capitol Hill are sitting on their hands at this critical time. They feel too beholden to Silicon Valley's lobbying and vote-getting machine to take any meaningful action—even when such action is obviously overdue.

But there are thoughtful outliers, including Senator Mark Warner of Virginia, who has put up a fix-it plan, and some of his ideas are worth consideration.

Warner wants to treat Facebook like a tightly regulated public utility. Specifically, he proposes:[12]

1. **Forcing major platforms to make activity data publicly available to researchers, while ensuring users' anonymity and privacy.**

2. **Making disclosure requirements for online political ads more akin to those for radio and TV ads.**

3. **Placing a kind of fiduciary responsibility on the social platforms, similar to those placed on the financial world.**

4. **Assigning privacy rule-making authority to the Federal Trade Commission.**

5. **Passing privacy protection and algorithmic fairness regulations similar to Europe's new GDPR regulations.**

Warner has also put forward some optional ideas, such as making platforms liable for claims such as "defamation, invasion of privacy, false light, and public disclosure of private facts."[13]

Now this would be a radical change, basically turning the platforms into the news services that they have *de facto* become, but insisted on not being labeled. Such a draconian move would either kill the platforms, or change them entirely.

On paper, or at least in theory, Warner's proposals look like fine steps taken in the right direction. But the devil will be, as it always is, in the details. How will these be enforced? Looking at the government's past attempts to protect consumers in the financial markets, you can see that the execution has been practically designed to fail.

So, once again, in the government's effort to protect consumers' data privacy, there will probably be multiple agencies involved, with no single enforcer. There will be overlapping authorities and jurisdictions. Years will pass with the only certainty being more buildings built in Washington to house more bureaucrats with little to show for the effort.

Federal Privacy Policy Working Meetings

In 2018, the President's National Economic Council held meetings to study and craft a consumer privacy policy seeking "the appropriate balance between privacy and prosperity."[14]

Over at the Federal Trade Commission, Chairman Joseph Simons has pledged "vigorous" antitrust enforcement hearings in late 2018. One sign the hearings are serious is the FTC's recent hiring of attorney Lina Khan. One of the brightest young minds on competitive markets, at Yale Law School she wrote a paper titled "Amazon's Antitrust Paradox" which brought "fresh thinking on digital platforms" by showing how "the current antitrust enforcement framework is ill-equipped to tackle Amazon's dominance" and, therefore, what new tools are need to pursue Amazon on antitrust grounds.[15]

Allow us some skepticism here.

Our experiences as observers of the government's ability to follow through on piecemeal attempts to enact broad federal privacy legislation support our view that anything less than a full-on effort isn't likely to slow down the social platform steamroller. Trying to craft compromises that "everyone can live with" will, as with so many previous such efforts, likely be rolled over and soon forgotten.

Left in place will be the social machinery that attempted to sway the 2016 and 2018 elections. And when Donald Trump seeks reelection in 2020, the platforms can be counted on to try every algorithmic trick at their disposal inspired and energized by their hatred of the president.

There are literally a quintillion reasons to take action against the social platforms. That's how much data we are generating—quintillions of bytes every day; a number followed by thirty zeros. A number that's doubling in the world every two years.

Multiplied by…

- The billions of devices that collect and transmit data.
- The massive data farms that make it easy to store unlimited quantities of data.
- The escalation to 5G bandwidth that can zip data faster than an eye blink.
- Crazy smart algorithms that can extract actionable information from mounds of chaotic data.

- Network effects with each new person connected to the platforms giving those platforms an exponentially greater influence and power.

Quintillions of bits of data flying about—all controlled by a handful of men. Reform proposals will, at best, only aim at portions of the problem. Data leaks, bots run amok, ads that track us online, data brokers, political ad disclosure, etc. And those are only problems we know about; we don't know what we don't know—but you can be certain there are many more problems lurking behind the Big Tech Tyrants' high walls. Even if we can only take on a few problems at a time, such a piecemeal approach essentially doubles down on current US privacy rules. We've seen how a system of stronger rules and occasional fines is no match to the Silicon Valley steamroller.

It is time for a more ambitious approach. A time to think bigger.

Big Tech Tyrants Try to Sue for Peace

By July 2018 it was clear the Big Tech Tyrants could no longer evade some kind of punishment. They had crossed too many lines too many times.

Unable to avoid an outright comeuppance and kneeling contrition, they began bargaining in earnest for a preemptive end to hostilities—in hopes of living to fight again, perhaps even resuming with business close to as usual.

- Mark Zuckerberg hinted in congressional testimony and during a media tour that Facebook should be regulated in some way.
- Apple's Tim Cook gave several interviews saying that self-regulation is no longer a viable option.
- All the social platforms are putting controls in place—to actually limit their use!

It was rich, watching the Big Tech Tyrants reckoning with their products, which were only doing precisely what they had been built to do: keeping users clicking and tapping and scrolling for hours and hours. And lest anyone try to tear themselves away from the dopamine drip, they were yanked back in with an incessant stream of notifications.

But to assuage Congress, the Tech Tyrants have all announced new tools that can limit time spent on the apps, monitor usage patterns and suggest healthier approaches, keep users from checking notifications at bedtime, and similar "helps."

Google even had the audacity to call their controls a "Digital Wellbeing" initiative. It's worth noting that their own design ethicist, Tristan Harris, had proposed these ideas five years ago but was ignored.

In a flurry of announcements and TV ad campaigns, Facebook insisted that...

1. **Somebody else caused the problem, but we'll fix it.**

2. **Somebody else meddled in the elections on our platform, but we'll fix it.**

3. **Okay, yes, we gathered external data on our users so we could get into their heads, but we'll fix it.**

This third announcement is probably the most revealing of whether Facebook means anything it says.

To wit, in July 2018, Facebook said it would cease gathering external data on users and shut down its "Partner Categories" tool. They had used external data to build more detailed personal profiles on users and better target ads at them, based on things like offline purchasing history.

This means companies advertising on Facebook will only be able to use their own in-house data and the data Facebook collects itself. This also means that Facebook could hemorrhage hundreds of millions in advertising revenues, and even trigger shareholder lawsuits for not seeking to maximize returns.

Think Facebook is going to do this voluntarily, and not revert to the old practices the first chance it gets? Advertisers have already hinted that Facebook will find a workaround. Heineken, for one, has said they are "optimistic" Facebook will find a solution to resume targeting once all the media attention dies down.[16]

But now, with new controls in place, controls intended to prove the Big Tech Tyrants really mean they will play nice this time, they've thrown their arms high over their heads in a show of surrendering to government investigators, concluding that they will fare best by making a show of submitting to regulations of some kind.

But what, then, when the media glare turns elsewhere—as it invariably does? Then the social platforms will again dispatch their legions of lobbyists across Washington to quietly water down any regulations they can't ignore, and loudly applaud those that really don't matter to them.

Time will pass and the Tyrants, who know this script well, will begin rehabbing their good names among the important decision-makers and customers, as well.

They'll trot out a who's who of business experts who don't believe the platforms should be throttled. Experts like Dr. Kim Wang, professor of strategy and international business at Suffolk University's Sawyer Business School, who has delighted in telling the *Wall Street Journal*: "Today's Amazon is tomorrow's Macy's. Very few companies will be able to position themselves for the new, next technology every time."[17]

The Tyrants will also argue that they're not anti-competitive.; they'll say "an alternative is always only a mouse click away."

Which is technically correct.

And they'll say, hey, if you don't like the platform, you're free to leave.

Correct again.

Oh, and remember the 1998 *Fortune* article declaring Yahoo! the winner of the search engine wars?[18] Google would launch a mere six months later—so we aren't the ogre monopolies, some say. We are always fighting for our lives.

Correct, to a point. Back in 1998 we were in the early stages of a hundred-year shift in technologies. Fewer than one in four Americans had even used a search engine. There were hundreds of companies vying to win this emerging new category called "search." Yahoo was only three years old itself—a puppy. *Fortune* was rushing to judgment, in hindsight. But twenty years later, Google's very mature control of search is entirely different, concentrated, and controlling.

All of these arguments will be made and made beautifully by some of the best storytellers for hire. What's more, the Tech Tyrants will surely marshal their user bases to their defense. Fact is, lots of people just don't care about personal privacy, political censorship, or addiction dangers. They just want their digital fix—be it a free spreadsheet or a free outreach to distant friends. These are valuable things to a great many people. No denying that. So the social platforms will raise up this user goodwill like a dyke before the approaching storm.

Meanwhile, battalions of attorneys will march into play, all armed with a single message: "Baby steps, people. Baby steps. Let's not do anything drastic."

The lobbyists and attorneys will argue that if the social platforms are broken up or taken down in one fell swoop, all hell will break loose. All of the tightly integrated underlying code bases will unravel and fall apart. It could be a death sentence, they will insist.[19]

We suppose that if you tried to break up the mob, they would marshal the same defense. We also suppose that if the social platforms applied even half the brilliance to fixing their code as they did to writing it originally, they'd be fine. These are, after all, very smart guys.

Very smart guys at social manipulation, but not socially smart guys.

For the most part lacking the basic social skills of their friends in the humanities, these single-minded engineers pushed ahead and built impressive social platforms on the water's edge of soft-sand beaches (ironically, the source of silicon). Now the tide of public outrage has come in, and we've seen just how destabilizing those platforms have become—taking a huge toll on society.

Now it is time to rebuild those platforms within the greater good of the American ideal.

The Common Sense Case for Breaking 'Em Up

With Amazon it comes down to a single question: How comfortable are we with a single company controlling 50 percent of online commerce and executing on a plan to control 30 percent of all commerce?

Today, being an Amazon Prime customer means lower prices and higher convenience…usually. What's not to like? It's all punch and roses until Amazon succeeds in its master plan to remove all of its many competitors. What then? It has been backed by investors who took losses for a decade on a bet that they could make oversized fortunes someday. Those investors will see their gamble pay off when Amazon can begin ratcheting up prices higher than we ever saw before. And Amazon can do it, because it no longer fears competition. It has taken care of that.

The fewer competitors, the more allowance Amazon has to control prices. Those who love Amazon now will feel very differently when Amazon is the only place you can shop for the things you need—and therefore it can, and historically will, charge whatever it wants.

Amazon Pulled Off One of History's Great (Anti)Capitalist Cons

Amazon's Jeff Bezos was first to see the potential of the internet to generally restrain trade (realizing this epiphany back when most folks still thought of the World Wide Web as some kind of magical "futureworld").

Bezos pushed at the constraints of legality with a three-part strategy:

1. Sell at below cost and subsidize losses with (a) investors committed to a long-term strategy and helped along by low-cost capital, and (b) other company divisions that were profitable. Amazon tapped the receipts of its cloud services division—a cash cow unrelated to retailing. This activity was definitely related to predatory pricing—an illegal setting of prices so low that they are aimed at eliminating the competition.

2. Avoid having to collect sales tax, offering customers the equivalent of up to 10 percent in discounts right out of the gate. To put that in perspective, over the past decade Amazon paid only $1 billion in taxes and Walmart paid $64 billion.[20] Sure, Walmart still books three times the sales of Amazon, but not sixty-four times the sales![21] That's a huge advantage for Amazon—created not by business acumen, but by legislative fiat.

3. Use a package delivery strategy that effectively co-opts the US Postal Service, along with FedEx and UPS, so that those services deliver to you a parcel from Amazon for a fraction of the price the same services charge their "regular" customers. USPS spokespeople say they cannot reveal how much Amazon actually pays the Postal Service, but that they are forbidden from Congress to lose money on any contract. What's more, they say, they can charge Amazon less than the going rate because the postal trucks are already on the route to people's houses, and Amazon's packages are just a marginal add-on. While correct from a bureaucrat's point of view, it's faulty logic any other way. It suggests that every business could conceivably pay less for postage because the postal truck is already on route. Who, then, would pay for anything? What's more, a separate Citigroup study found the USPS was charging an average of $1.46 below market rates for parcel delivery.[22] And Amazon's billions of packages are contributing to those losses, while saving the company billions of dollars that others have to pay.

The longer Amazon is allowed to game capitalism with this three-part strategy, the more potentially dangerous it becomes, not just for customers, but for the very retail ecosystem Amazon is creating with its suppliers, competitors, and workers alike.

Amazon Is Soon Fully Vertically Integrated

It sells other companies' products while manufacturing and selling its own goods as well. All of them on its own platform. In fact, Amazon actually sees a competitor's prices before market, before potential buyers do, allowing Amazon to know which products are doing well. Armed with that insider knowledge, they have the keys to both product construction and competitors' price points, allowing them to make and sell their own knockoff products at a lower price, right on the same page.

That's what Amazon did to Quidsi, owner of online shopping stores such as Diapers.com.[23] Amazon had tried to buy Quidsi in 2009, but the company wanted to stand on its own. Not to be outdone, Amazon began undercutting Quidsi's prices for diapers and other baby products by nickels and dimes through a new service, Amazon Mom. With sales plummeting, Quidsi was forced to agree to an acquisition. It wasn't long before Amazon shuttered Diapers.com, having eliminated the competition.

This kind of trading on inside information falls into a legal gray area.

When business laws were created decades ago, it wasn't possible to see into a competitor's pricing plans without breaking into its offices. Now on the platforms, it's easily done—it's even an automated process—and the laws have a lot of catching up to restore a level playing field.

In the view of a former director at the Federal Trade Commission, Tom Campbell:

> Amazon has become such a preferred location, a preferred online shopping space, that if Amazon is competing with somebody else, that somebody else nevertheless has to advertise on Amazon, and so Amazon is in the position to find out its competitors' pricing and choose to undercut the pricing if they wish.[24]

Antitrust law has traditionally attempted to protect consumers from predatory business practices while ensuring an open market that fosters fair competition. But Amazon is effectively sending this message to potential competitors:

> "We have very deep pockets, and we can charge below break-even prices for years if need be—until you've been bled of every dime in your

company. And so, it may not be worth it for you to even enter this market. We'll win every time out, and you'll pay dearly for your foolishness."

Legally, predatory pricing cases rely on a two-part equation

Simply saying that Amazon is big, that it charges low prices, that it has driven thousands of small retailers into bankruptcy, is barely *prima facie* evidence of consumer harm. It's very difficult for a plaintiff to make a successful predatory pricing case. However, there are gray areas.

Randy Stutz is an expert at the American Antitrust Institute, and he believes Amazon is vulnerable because of its ability to snoop on its competitors selling on its platform: "There's a question as to whether that's, on some level, exclusionary or, on another level, it's just hard competition that arguably benefits consumers."[25]

This leads to bigger questions in antitrust law:

1. **Should a company's behavior toward consumers be the gold standard?**

2. **Or should government take a different approach in a platformed era?**

By 2020, Amazon is expected to control 10 percent of all retail sales in the country on its way to 30 percent—this is far from a monopolist's share. And, so, under current laws, Amazon is safe from being branded a monopolist. Yet only under current law. Congress is always, and should always be, enacting new laws to meet changing needs. And the Supreme Court is always looking at new cases and being asked to reinterpret legal precedents over what is effectively a monopoly.

President Trump has called out Amazon's monopolist behavior

Wearing his leftist politics on his sleeve, Bezos has long clashed with Trump. In behavior less dignified than one would appreciate in any CEO, much less one who aspires to control retail in America, Bezos once tweeted that he would like to send Trump into space on one of his rockets.[26]

Whether joking or jousting, this is a man who may have started out with a garage operation, but has come to think of himself as the most powerful new resident of Washington, D.C.

Time to Bring Common Carriage Law to Bear?

When Twitter was caught shadow banning conservative publishers, and doing the same to Republicans in Congress in July 2018, Trump took to Twitter to denounce shadow banning, vowing to "look into this discriminatory and illegal practice at once!"

Trump's position can be supported by a consideration of centuries of common law precedent. Going back to the 17th century before we were a nation, the United States of America has had what are known as "common carriage" laws to protect people from discrimination at shipping ports and ferries, among other points of public transfer. Over time these protections were extended to railroads and telegraphs, and then to air travel. These protections were eventually codified into enforceable regulations.

Now, with Twitter discriminating at its whim, its execs insist they are within their right as a private platform owner, and they have claimed full exemption from any kind of common carrier restraint.

But are they…going forward?

Should Twitter and its similar competitors or spinoffs be classified as the new public points of transfer and subject to common carrier rules…and therefore no longer be able to discriminate at whim?

Like Twitter, Facebook has moved beyond its original moorings as a private platform governed solely by the Digital Millennium Copyright Act of 1998. Facebook now regularly makes editorial choices, as well. Facebook's News Feed algorithms and human curators decide which content will be seen and which will be banned, making them a point of transfer of critical public data over which they are employing a chokehold.

No longer just a platform but, in fact, a publisher, as evidenced by the actions Facebook is taking to control information on their platform. However, their policing actions have been shown to result in arbitrary and deliberate censorship.

In the olden days, this kind of problem, when it popped up, could eventually be resolved in the public interest through a time-tested mechanism: market competition. But Facebook is a monopoly platform/publisher/carrier. Market mechanisms no longer exist.

However, with common carriage protections applied to these new technologies, like they've been applied to past technologies, the problems of shadow banning and outright discrimination will have an equitable means of resolution.

Facebook Has an Actual App for Restraining Trade

Facebook, not satisfied with becoming merely anti-competitive, had purchased a security app to make themselves *even more* anti-competitive. The app, called Onavo Protect, would constantly be crawling the internet collecting information on nearly everything Facebook users do online: what they download, where they visit, what apps they use, everything.[27]

This gives Facebook a real-time log of all the new products coming out, including which products are gaining in popularity enough to pose a competitive threat. Facebook then moves to acquire the company, like it did with the WhatsApp messaging tool. Or Facebook develops a competing product, like it did with live-video streaming.

Facebook is the leadoff hitter in what are often called the FAANG companies—for Facebook, Apple, Amazon, Netflix, and Google. The acronym does have a certain ring and so it has caught on. But it doesn't begin to capture what's going on. Apple and Netflix are driven by different business models, and Twitter is left out. But FAAGT—for Facebook, Apple, Amazon, Google, and Twitter—doesn't readily roll off the tongue. FATGASP—for Facebook, Apple, Twitter, Google, Amazon, and Social Platforms—has a ring, of sorts, but still falls short. Perhaps the best capture of the social platforms' misbehavior has been offered by the editors at *The Economist*:[28]

"BAADD." Big, Anti-Competitive, Addictive, and Destructive to Democracy

Echoing these charges is Scott Cleland, former deputy US coordinator for communications and information policy in the Bush Administration, and author of *Search & Destroy: Why You Can't Trust Google Inc*:

Information is power; power corrupts, and absolute power corrupts absolutely. It's hard to imagine a more harmful monopoly in a free market democracy than one with bottleneck control over the supply and demand for accessing the world's digital information. That's because the free flow of information and the competition of ideas are the lifeblood of both constitutional democracy and free market competition.[29]

Asked for proof of the direct harm of this bottleneck to a competitive marketplace, Cleland gave some hard numbers on the harm to app makers:

> Google required all device manufacturers and mobile carriers, that used Google's free Android operating system, to make Google Search the default search engine and to prominently pre-install and prioritize over a dozen Google apps. Now 16 of the top 20 Android apps are Google apps.

Google's business practices allowed it to blindside the entire app-making community with such monopolistic power that only four of the world's twenty most popular apps survived. Google pummeled them into submission by front-running their own apps.

"Front-running" is quite illegal in stock trading and is obviously a relevant consideration when looking at the behavior of these social platforms.

This was not the first time Google and antitrust appeared in the same sentence. In August 2012, the professional staff of the Federal Trade Commission leveled five counts of antitrust violation against Google, and recommended prosecution. What happened next speaks volumes.

In November of that year, newly re-elected President Obama appointed Renata Hesse as acting assistant attorney general for antitrust. Hesse's previous job had been as an antitrust lawyer for Google. Just forty-four days into her tenure at the Department of Justice, all five Google antitrust probes magically went away.[30]

Two years later, Google's cozy relationship—some would say undue influence—with the Obama Administration again surfaced. Google was facing multiple patent infringement lawsuits from Apple, Microsoft, Nokia, Black-Berry, Ericsson, Sony, and Skyhook Wireless.

It was well-known Google had infringed, and their defense was basically, "Nah, nah, na, hah, nah." And indeed, a woman named Michelle Lee was soon appointed acting director of the US Patent and Trademark Office—in charge of developing the case against Google. Just eighteen months prior, she had been Google's deputy general counsel for patents. You won't be surprised that within months of Lee's taking over the patent office, all the companies suing Google settled or dropped their cases.[31]

Niche Social Networks Prove Breaking Up Works

Specific-interest apps are growing in popularity as many people tire of the big, noisy, sprawling platforms. There's Runkeeper for serious runners, with some forty-five million users; AllTrails for hiking and camping, with seven million users; Untappd for beer drinkers, with three million users.

Two communities in earlier stages of development are FeedMe (feedme. app) and USA.Life—both hoping to provide an alternative to Christians and conservatives who are being censored on Facebook or Twitter, but are really open to anyone from any persuasion. Their goal is to give the user the control of the content they see in their feeds and not the platform itself using their algorithms.

Speaking to the attraction of these kinds of niche platforms, professor Scott Campbell of the University of Michigan says:

> "People are sick of algorithms, sick of News Feed manipulation, in some cases sick of mood manipulation and really worried about trolling and bots on Facebook and on Twitter, and those things are easier managed and controlled in a smaller network. I also think with privacy, people are just not trusting, especially of Facebook, anymore. They lost trust."[32]

These smaller, niche platforms will never be a threat to Facebook in its current form. But they are at least succeeding at being "lean-in" experiences—to borrow Facebook COO Sheryl Sandberg's book title, ironically—since Facebook has increasingly become a "lean-back" experience where folks passively "lurk and like" to pass away the hours, getting just enough dopamine hits to keep at it.

These niche platforms also testify to the wisdom of busting up Facebook into multiple companies. If, instead of one Goliath, there were lots of "mini-books" competing with each other, the market could, and presumably would, punish poor data privacy or political banning. When enough folks can move to a competitor—in a leveled marketplace—a market message could be sent and received. All of the mini-books could work to create their own kind of socially fulfilling experience, advancing the technology in the best interests of society.

In practice, such a breakup of a tech monopoly has precedent in the dissolution of AT&T in 1982 and the creation of the "Baby Bells" and other telecommunications companies (and, eventually, ISPs) during the 1980s and 1990s.

The economic outcome of the AT&T breakup did not bear the hoped-for fruit of lower rates, but it did create a model for future consideration of monopoly-breaking. Whether such a break-up matrix could be applied to

FANG/BAADD and others is fraught with problems, not the least of which could be political upheaval on a national scale, but isn't it worth considering given the steamrolling alternative headed our way?

Last Time These Tyrants Can Be Stopped?

The need for aggressive action becomes even more pressing with the social platforms pushing into the AI era. AI feeds on massive amounts of data to deliver the powerful results of which it's capable. All of the Tech Tyrants are pushing AI's envelope, looking for more ways to use machine-learning algorithms in their operations.

What's more, they are putting AI to work in virtually all of the industries they're buying up or exploring—from driverless cars to software that decides whether or not you should get a loan. This will impact each of our lives very directly—from all the jobs that will be lost, to the inability of other firms to catch up. **If no constraints are placed on these Big Tech Tyrants, they will soon be in control of every tool that defines our national course.**

For a quick word picture of how these Tyrants see our future unfolding, and how right they believe themselves to be, here's a story from Jonathan Taplin—author of *Move Fast and Break Things: How Facebook, Google, and Amazon Cornered Culture and Undermined Democracy.*

It happened in May 2018. Google had gathered up all their engineers to unveil a new technology—the Duplex digital assistant. In an onstage demo, Duplex phoned a hair salon and made an appointment to get its hair done. Duplex did its best to mimic real human speech, complete with the "ums" and "likes" so many people use. Over at the hair salon, the receptionist had no idea she was talking to a bot. All the engineers whooped it up.

Taplin had this insight:

> If ever there were a metaphor for how clueless Big Tech is about the notion of privacy, and technology's growing role in our artificial-intelligence-mediated world, this was it. That the crowd of programmers was pleased by the idea that an AI assistant could fool a human should give us pause.[33]

We are being "fooled" on several fronts.

1. **Fooled into handing control of our most personal secrets to the Tech Tyrants.**

2. Fooled into thinking that these Tech Tyrants have our best interests in mind.

3. Fooled into accepting addictions' hammer, Stasi-like tactics, and thought manipulation as fair exchange.

How Tyranny Begins

True tyranny, the kind America has never really known but which is stinking up things now, has always begun with a powerful authoritarian censoring of free speech and ended with a single point of view imprinted on everyone else.

When what you say is shouted down or banned, or when those with whom you agree and want to listen to are blocked from reaching you, that is authoritarian, that is oppression—that is a Silicon Curtain dropping down between those who wish to speak freely, and those who want to listen to free speech.

The Tyrants are pulling that Silicon Curtain's cord—Becoming a "Totalitarian Place."

When people are unanimously on one side, that tells you not that they've all figured out the truth but that they're in a sort of totalitarian place, that they're in a one-party state, where they're not allowed to have dissenting views. I think somehow Silicon Valley shifted from being quite liberal to being a one-party state. Those are two very different things.[34]

—PETER THIEL, VENTURE CAPITALIST
AND TRUMP SUPPORTER

In the our past, we only concerned ourselves with an individual being the only voice in the "public square." However, there was usually the chance of moving to a different "square" or finding a different soapbox around the corner. However, when the dominant voice not only shouts the others down but builds itself into the only public square in the game…tyranny is emboldened by their monopoly.

More Accurately, Big Tech Tyrants Are Predator Monopsonists

If ever there was a word that couldn't get "friended" on Facebook, but describes the social platforms' behavior precisely, it is "monopsony." Even economists shy from using the word monopsony—and it's hard to blame them. Everyone knows what a monopoly is, or, at least, you know it when you see it. But trying to wrap your mind around monopsony feels like it'll fry a lot of brain cells. But trying anyway, since it matters…

A monopoly is to a hammer
as a monopsony is to a funnel.

We think of a *monopoly* as a company controlling the supply of a product that people need, so the company can hammer people with whatever prices it wants.

A *monopsony* is when a company inserts itself between the sellers of a product and a single buyer, acting like a funnel so the company can control what passes through that funnel.

Amazon can influence the price that sellers charge on its platform and can even undercut those sellers with millisecond precision—so that Amazon products are the winning products in each new category Amazon enters.

Facebook, Google, and Twitter can influence who has access to the billions of people looking to shop for things they want, learn what's going on in the world, and engage in politics. For example, when looking at how the Tyrants censor the delivery of information, it's the same as if the Post Office chose to deliver some—but not all of your mail.

Apple, however, is an exception, and it's not fair or accurate to lump Apple in with these new monopsonists. Apple's business model will always be vulnerable to competitive disruption—despite the fact that it currently tops the tech world in revenue, profit, market capitalization, and cool.

> **Today's monopsonists differ from the monopolists of robber baron fame in that they learned one thing from those robber barons: Whatever you do, don't go turning armed guards on striking workers. Better to ship the jobs to distant ports where nobody can see the slave labor shops that make your product. Better still to automate away all of those pesky human jobs and let someone else worry about a future with anything but menial jobs. Better to use your clever algorithms to advance your political agendas, and then feign trustworthiness. And when found out, better to retreat into superbly scripted TED Talks all about elevating awesomeness to awesome new levels.**

Figuring Out Antitrust Law for a Platformed World

Of all the insights of enlightened western culture, we most prize having a government of laws, not of men (or persons, contemporizing). All else has flowed from this insight, attributed to the 17th century political theorist James Harrington and popularized by John Adams in the 1780 Massachusetts state constitution. Among the most important of these laws were those related to our first freedoms—our privacy. As Americans, we held these freedoms dear. And there are men and women in uniform today who are representing these freedoms—and dying for them, even as those freedoms come under attack here at home.

First to usher these freedoms into the industrial age was Louis Brandeis, later one of our great Supreme Court justices. In an 1890 Harvard Law Review article, Brandeis laid out case law and statutes for citizen privacy which he called being able "to be let alone."[35]

This law held until the arrival of mainframe computers, which made obvious the need for updated statutes. So in the 1970s, we adopted federal data privacy laws such as the Fair Credit Reporting Act and the Privacy Act to control how big institutions and government could use our personal information.

Two decades later the internet was taking off, and Congress sought to tighten privacy protections on two important fronts. The Children's Online Privacy Protection Act made it illegal for businesses to collect personal information from children under age thirteen. And the Digital Millennium Copyright Act sought to protect artists from theft of their work (it didn't) and offer internet companies blanket immunity from anything posted on their platforms (it did).

Into the new millennium, with companies now able to collect, aggregate, and correlate data at lightning-fast speeds around the clock, any privacies ensured by these previous statutes went out the window.

- It became possible to identify any individual in America by linking up sources of anonymous information on her.
- It became possible to know every little thing about her—to know what she's doing now, what she'll buy next, where she's going tomorrow (even, as we pointed out earlier, when she may die) and to profit off of this information even if it harms her.

No laws or regulations address this new reality. We've gone as far from Brandeis' idea of being let alone as a zero is from a googol (ten to the hundredth power, which few knew until some Stanford students mistyped it, got "google," and ran with it).

Today, every step we take, every move we make, someone's watching us, like the stalker in The Police song. We've gone so far, and so fast, we are often deemed to have no expectation of privacy. It's even said, and often, that privacy is dead. Sun Microsystems CEO Scott McNealy said, "You have zero privacy, get over it."

Moreover, so much of this data collection is now automated that it'd be almost impossible to regain any measure of real privacy, shy of pulling off Tom Hanks's *Cast Away* act.

Recall, how *Wall Street Journal* writer Julia Angwin told of her off-the-digital-grid experiment in which she spent a year trying to live her life without leaving a single digital trace. As she concluded in her book, *Dragnet Nation:* "The average citizen stands little chance at protecting basic privacy, and we're paying a bigger price than we know for the privacy we've forfeited."

As for the notion of "informed consent" that was plausible two decades ago, that's only a fantasy today with the constant stream of online interactions we have, with the football-field-long privacy policies and terms and conditions and user settings we're supposed to keep on top of.

We're inundated. But nowhere is it carved in stone that we've moved beyond the reasonable expectation of privacy that previous generations of Americans enjoyed—based solely on our living at the end of the second decade of the second millennium.

Trust-Busting: Forgotten Formula for Innovation and Progress

Some say we should bust up today's Tech Tyrants in the same way they busted up oil and railroad monopolies a century ago.

Others insist that today's technologists are guilty only of delighting people. And should they fail at that, the market will surely bring them down a peg, as it has always done.

At the root of this debate is a fundamental question:

When does a company cast such a large shadow over the nation's economy that it extinguishes all competition—no matter how happy the people are with its products?

In the early 1900s, Teddy Roosevelt believed Standard Oil had crossed that line. There's little doubt that Standard Oil was the Facebook, Google, and Amazon of its day.

John D. Rockefeller had given the nation a technological marvel. Prior to his engineering advances in refining oil, the average home was lit with expensive whale oil, which few could afford. So yes, everyone initially loved the man! But as time passed, Rockefeller used his marvel to ruthlessly drive any competition to an early grave. As the muckraking journalist Ida Tarbell wrote of Standard Oil: "They fought their way to control by rebate and drawback, bribe and blackmail, espionage and price cutting, and perhaps more important, by ruthless, never slothful efficiency of organization."[36]

So it was Standard Oil's behavior, not its product, that prompted Roosevelt to sue Standard Oil for antitrust violations. From the outset, the government's lawyers were well aware that they didn't have an open-and-shut case.

So the government made the case the nation needed to have made: that when a company threatens to overpower all comers, there must be rules to give other companies a fighting chance. That's what happened. Standard Oil was split into thirty-four firms.

This became the standard for government antitrust enforcement for decades to follow. Any time a company's technological advances allowed it to undermine competitors—as would happen with Alcoa, AT&T, IBM, Kodak, and others—the government would step in.

The key factor was technological advantage. When that advantage turned a company into a monopoly, government intervention became necessary. And as a corollary benefit—though few, if any, knew it back then—the very act of intervening actually created a cascade of additional benefits to society that were not at once obvious.

In the 1980s, for example, IBM ruled over the computer world, and an antitrust case was brought. This, in turn, opened the door for a young company named Microsoft to run through. By the 2000s, it was Microsoft ruling over the computer world and an antitrust case was brought. Microsoft appealed and won, but agreed to break into two separate units—operating systems and software products. This opened the door for competing browsers such as Netscape and Firefox and, later, because Microsoft's monopoly had been weakened, it opened the door as well to young companies named Google and Facebook.

What antitrust regulators learned over this time is that innovation and regulation are a cyclical pair, the one requires the other for society to advance.

Clearly, we are once again at this inflection point between innovation and regulation.

The social platforms realize it, as well. They are backpedaling as fast as possible. Trying to deflect those criticisms that they can't get out ahead of. Promising to fix a long list of problems and addressing all the complaints against them. But they cannot change the essential equation:

- Their business models are predicated on violating privacy.
- They are committed to censoring all those they disagree with.
- They are now monopsony powers and defying the common good.
- Progress comes from opening up the field to competitive innovation.

Some Ways to Make the Game Right

1. We have seen Silicon Valley technologists take control of our data—trampling basic privacy with addictive algorithms.
2. We have seen them concentrate their markets—eliminating any possible competition for decades to come.
3. And we have seen them exerting their political views—turning past elections and threatening coming ones.
4. Too many lines have been crossed.
5. So, just as early in the 20th century, Teddy Roosevelt broke up Standard Oil, early in the 21st century, Donald Trump should break up the Big Tech Tyrants.

Options to Consider Going Forward

Crack Open the Black Boxes of the Tech Tyrants
Technologists running these social platforms will claim their proprietary algorithms must be kept confidential. But these algorithms are being *knowingly* used to manipulate the activities and thoughts of millions of *unknowing* people. The basic tenets of business trust are being violated—repeatedly. So, the veil on their operations must be lifted; their black boxes opened to the public eye.

These technologists have sought to become the operating systems of our lives—and they have largely succeeded. Yet we know little about how they really work. We must know.

Move the Tech Tyrants into the Public Commons

Social platforms could quite plausibly be declared "common carriage" entities and thus subject to all the liability, discrimination, censorship, and copyright restrictions that other public services are held to—hewing to a longstanding principle of our nation that the fair and proper functioning of these services is in the essential best interest of society. As we saw earlier, the Supreme Court has ruled against the "company town" as an arbiter of laws—that precedent should apply to the Tech Tyrants as well.

Rewrite the Communications Decency Act

Section 230 of the Communications Decency Act shields the social platforms from any liability for the content their users produce and host. This act helped launch the internet, but it may be time to revisit its value to our nation now.

Slap Meaningful Fines on the Tech Tyrants

Direct the Federal Trade Commission to investigate the Tech Tyrants and issue sizeable fines when/if they are determined to be censoring political voices they happen to disagree with.

We know the Tyrants have been threatened with fines in the past, including the FTC's 2011 consent decree which could result, if enforced to the letter of the decree, of fines to Facebook of $72 trillion. Enforce the law; slap 'em.

Bring a Special Prosecutor to Examine the Tech Tyrants

Appoint a special prosecutor to investigate the social platforms for their interference in the 2016 and 2018 elections, and their plans for 2020. The investigation should seek to determine whether there has been a "conspiracy to defraud the United States of America."

This mirrors the broad and nonspecific language that empowered Special Prosecutor Robert Mueller to cast an overwide net into Russian interference in the 2016 election, which thus far has brought indictments for everything but examining Big Tech's interference. Since this has been a prosecutorial model praised by progressives in Silicon Valley, they should welcome it being applied to them, as well.

And all should agree that if the social platforms were themselves interfering in the US elections, that is a far more damaging indictment. Put differently...

1. By silencing conservative voices in a coordinated campaign, weren't Tech Tyrants actually making illegal campaign contributions to liberal candidates?

2. Weren't they attempting to "steal" the 2016 and 2018 elections?

3. And what's to stop them from attempting to turn the 2020 elections as well, as they have said they wish to?

Require Tech Tyrants to Charge Competitive Fees

Social platforms could be required to charge fair market rates for their services—turning free subscribers into paid subscribers. However, in fairness, the entire history of the internet has demonstrated the difficulty in getting people to pay for things in the consumer realm. Video and music sites have enjoyed some limited success with this model—which explains the ascendency of Netflix and HBO. But otherwise it has not worked well. It is probably an unfair way to fix an untoward business model—but it is an option.

Each of these options—cracking open the black boxes, slapping meaningful fines, revisiting the Communications Decency Act, bringing in a special prosecutor, forcing a change in business models—can play a role in the larger goal of breaking up the Tech Tyrants to restore free and flourishing competitive markets to America.

Additionally, here is an outline of some plans to consider...

Break-'Em-Up Plan for Facebook

Three key objectives:

1. Separate Facebook and the companies it acquired to concentrate power.

2. Open Facebook's data stores to competing businesses.

3. Mandate stronger privacy rules to protect remaining Facebook users.

Facebook ideally becomes four different companies...

1. Facebook
2. Messenger
3. WhatsApp
4. Instagram

Four separately owned companies. Each with a different core strategy. This should reignite competitive innovation. It should diffuse the concentrated control of online conversation. It should leave each new firm with sufficient size and market share, so they can compete and thrive, if well-managed.

Break-'Em-Up Plan for Google

It is ironic that a company that first unleashed tidal waves of human creativity and learning is now concentrating power and distorting the marketplace, squashing competition and any innovation an upstart might attempt, while advancing their own narrow political agenda and stealing away basic privacies. But it is true.

There are several ways to break up Google's parent company Alphabet, including:

- Separate out Google search, and open up search advertising to other vendors.
- Divide Google's and YouTube's advertising units into standalone companies.
- Spin off as standalone companies the advertising units—formerly DoubleClick and AdMob, but recently renamed and consolidated onto a single ad platform.
- Create standalone companies out of YouTube, Android, and Google's cloud services (Gmail, cloud storage, maps, etc.), separating all of them from Google search.
- Regulate Google like a public utility, forcing it to license out its algorithms to help spur competition (like the government did with AT&T in 1956).
- Forbid Google from acquiring additional tech companies like Spotify or Snapchat.
- Alphabet reaches into many areas—email, thermostats, phones, driverless cars, AI, and virtual reality. They can spend money developing

new markets and blue-sky technology projects because 80 percent of their revenues come from advertising. With advertising turned into a separate company, Alphabet's own innovation might be hurt as much as it hurts others. One option is to keep Alphabet's "Other Units" in the same new company as Search.

These are all options to consider, and if the goal is to both acknowledge Google's many achievements while also restoring innovation and free-flowing ideas to the marketplace, then ideally Google's parent company Alphabet becomes five different companies…

1. **Google Search**
2. **Cloud Services**
3. **Android**
4. **YouTube**
5. **Other Units**

Each of these five can be strong and vital in their niche, capable of competing and thriving on their own, but no longer so big they can choke off competition by acquiring any real competitor or putting them out of business by denying them true access to the online marketplace.

Break-'Em-Up Plan for Amazon

Amazon has become so dominant in retail—as it stands today and where's its headed—that Amazon should be made to choose: Either sell goods as a standalone retail platform, or run a digital platform other merchants use to reach customers. If it chooses to be a platform, it must spin off the Whole Foods supermarket chain it recently acquired.

Amazon could ideally operate as five different companies:

1. **Amazon Web Services**
2. **Amazon Logistics**
3. **Amazon Retail**
4. **Amazon Entertainment**
5. **Amazon Health & Life Sciences**

Break-'Em-Up Plans for Other Tech Tyrants

Beyond Facebook, Google, and Amazon, there are not, in our view, compelling reasons to break up any other Tech Tyrants.

Twitter has a corrupted business model as well, and it uses that model to attack its political opponents. But there is no easy way to break it up, and little to be gained from doing so. Society may benefit more if Twitter is *de jure* transformed into the common carriage entity that, *de facto*, it is.

Apple has earned its distinction as the first trillion-dollar company. It is not without its faults, but none that deserve any draconian interference from government. Instead, Apple should be simply subject to rigorous oversight to ensure that it does not use its privileged position as a chokepoint to political discussion, or to stifle views it may disagree with.

In reaching our own conclusions about what's in the best interest of America, we have tried not to forget how much we truly loved these social platforms in their early days. In our personal and in our business lives, they opened up all kinds of possibilities—and we will be forever grateful to their founders and their backers for all they've given us.

Yet clearly their early wonders have been overshadowed by monopolistic—technically monopsonistic—behavior that's squashing all competition and innovation, laying a big thick wet wool blanket over commerce.

As conservatives ourselves, we are always circumspect about government interventions in a market economy. We see little gain from the expansion of the administrative state into the social platforms. But we believe there are actions to be taken now, actions that are needed to restore the integrity of our liberal democracy underpinned by free-market enterprise.

If the Big Tech Tyrants are not checked now, America will inevitably lose its diversity of media voices as a single worldview is forced upon a nation in a systematic campaign toward uniform thought.

We know the social platforms played a significant role in the 2016 election, and the same behind-the-scenes string-pulling took place in the 2018 elections. We know that the leftist leadership of the social platforms aims to influence elections going forward and certainly in 2020—they have said as much publicly.

Though America has risen to the challenge of antitrust prosecution many times in our history, we have not in recent years. Whether or not that is due to the leftist hold on key congressional committees and executive offices no longer matters.

An opportunity exists now to strike a blow for our democracy, for the ideals we hold dear in our free and open society, and for the world our children will live in...

It's time to break 'em up!

Endnotes

STRIKE 1: ADDICTION ALGORITHMS

1 Antonio Martinez, *Chaos Monkeys* (New York: Harper Collins, 2016).

2 Scott Galloway, *The Four: The Hidden DNA of Amazon, Apple, Facebook and Google* (New York: Portfolio/Penguin, 2017).

3 Jonathan Freedland, "Aggression, Abuse and Addiction: We Need a Social Media Detox," *The Guardian*, Aug. 4, 2018, https://www.theguardian.com/commentisfree/2018/aug/04/social-media-detox-facebook-twitter-august.

4 Walter Kirn (@walterkirn), Twitter post, April 9, 2018, https://twitter.com/walterkirn/status/983338972322652160.

5 Marty Swant, "Facebook Is Building Its Own Neuroscience Center to Study Marketing," *Ad Week*, May 23, 2017, http://www.adweek.com/digital/facebook-is-building-its-own-neuroscience-center-to-study-marketing.

6 Jaron Lanier, *Ten Arguments for Deleting Your Social Media* (New York: Henry Holt & Company, 2018).

7 Justin Lafferty, "Facebook Ads to Become More Relevant, Say Why They're Being Shown," *Ad Week*, June 12, 2014, https://www.adweek.com/digital/facebook-ads-to-become-more-relevant-say-why-theyre-being-shown/.

8 Zeynep Tufekci, "Algorithmic Harms Beyond Facebook and Google: Emergent Challenges of Computational Agency," http://ctlj.colorado.edu/wp-content/uploads/2015/08/Tufekci-final.pdf.

9 Sally M. Gainsbury, Paul Delfabbro, Daniel L. King and Nerilee Hing, "An Exploratory Study of Gambling Operators' Use of Social Media and the Latent Messages Conveyed," *Journal of Gambling Studies* 32, no. 1 (March 2016), 125-141.

10 Katerina Eva Matsa and Elisa Shearer, "News Use Across Social Media Platforms 2018," Pew Research Center, September 10, 2018, https://www.journalism.org/2018/09/10/news-use-across-social-media-platforms-2018/.

11 Lanier, *Ten Arguments*.

12 Brian A. Primack, Ariel Shensa, Jaime E. Sidani, Erin O. Whaite, Liu yi Lin, Daniel Rosen, Jason B. Colditz, Ana Radovic, and Elizabeth Miller, "Social Media Use and Perceived Social Isolation Among Young Adults in the U.S.," *American Journal of Preventive Medicine* 53, no. 1 (July 2017), 1-8.

13 Christina Sagioglou and Tobias Greitemeyer, "Facebook's Emotional Consequences: Why Facebook Causes a Decrease in Mood and Why People Still Use It," *Computers in Human Behavior* 35 (June 2014), 359-363.

14 Ethan Kross, Philippe Verduyn, Emre Demiralp, Jiyoung Park, David Seungjae Lee, Natalie Lin, Holly Shablack, John Jonides, and Oscar Ybarra, "Facebook Use Predicts Declines in Subjective Well-Being in Young Adults," PLoS ONE 8(8).

15 Holly B. Shakya and Nicholas A. Christakis, "Association of Facebook Use With Compromised Well-Being: A Longitudinal Study," *American Journal of Epidemiology* 185, no. 3 (February 2017), 203-211

16 Robert Booth, "Facebook Reveals News Feed Experiment to Control Emotions," *The Guardian*, June 29, 2014, https://www.theguardian.com/technology/2014/jun/29/facebook-users-emotions-news-feeds.

17 Zeynep Tufekci, *Twitter and Tear Gas: The Power and Fragility of Networked Protest* (Boston: Yale University Press, 2017).

18 Lisa Wirthman, "How to (Legally) Use Social Media to Recruit," *ADP Brand-Voice*, published on *Forbes.com*, Oct. 24, 2016, https://www.forbes.com/sites/adp/2016/10/24/how-to-legally-use-social-media-to-recruit/#1fd4ebce29f4

19 Yanhao Wei, Pinar Yildirim, Christophe Van den Bulte, and Chrysanthos Dellarocas, "Credit Scoring With Social Network Data," *Marketing Science* 35, no. 2 (July 2014), 234-258.

20 Gina Roberts-Gray, "Car Insurance Companies Use Facebook for Claims Investigations," *Edmunds.com*, Sept. 4, 2013, https://www.edmunds.com/auto-insurance/car-insurance-companies-use-facebook-for-claims-investigations.html.

21 Yeganeh Torbati, "Trump Administration Approves Tougher Visa Vetting, Including Social Media Checks," Reuters, May 31, 2017, https://www.reuters.com/article/us-usa-immigration-visa/trump-administration-approves-tougher-visa-vetting-including-social-media-checks-idUSKBN18R3F8.

22 "Social Media Shocker: Twitter and Facebook Can Cost You a Scholarship or Admissions Offer," *Tuition.io*, April 24, 2014, https://www.tuition.io/2014/04/social-media-shocker-twitter-facebook-can-cost-scholarship-admissions-offer.

23 Sam Levin, "ACLU Finds Social Media Sites Gave Data to Company Tracking Black Protesters," *The Guardian*, Oct. 11, 2016, https://www.theguardian.com/technology/2016/oct/11/aclu-geofeedia-facebook-twitter-instagram-black-lives-matter.

24 T.S. Allen and Robert A. Heber Jr., "Where Posting Selfies on Facebook Can Get You Killed," *The Wall Street Journal*, July 26, 2018, https://www.wsj.com/articles/where-posting-selfies-on-facebook-can-get-you-killed-1532642302

25 Sevinc Kurt and Kelechi Kingsley Osueke, "The Effects of Color on the Moods of College Students," *SAGE*, January-March 2014: 1-12.

26 Lanier, *Ten Arguments*.

27 David Meyer, "'Deceived by Design:' Google and Facebook Accused of Manipulating Users Into Giving Up Their Data," *Fortune*, June 27, 2018, http://fortune.com/2018/06/27/google-facebook-dark-patterns-privacy/.

28 Wendy Davis, "Google and Facebook Push Users to Cede Privacy, Advocates Say," *Digital News Daily*, June 27, 2018, https://www.mediapost.com/publications/article/321424/google-and-facebook-push-users-to-cede-privacy-ad.html.

29 Ibid.

30 John M. Simpson, "Eight Consumer Advocacy Groups Call Upon FTC to Investigate Google and Facebook," *Consumer Watchdog*,

June 27, 2018, http://consumerwatchdog.org/privacy-technology/
eight-consumer-advocacy-groups-call-upon-ftc-investigate-google-and-facebook.

31 Christopher Mims, "Who Has More of Your Personal Data Than Facebook? Try
Google," *The Wall Street Journal*, April 22, 2018, https://www.wsj.com/articles/
who-has-more-of-your-personal-data-than-facebook-try-google-1524398401.

32 *60 Minutes*, "The Data Brokers," CBS News, March 9, 2014, https://www.cbsnews.
com/news/the-data-brokers-selling-your-personal-information/.

33 Mims, "Who Has More."

34 Richard Nieva, "Google Explains Gmail Privacy After Controversy," CNet, July 3,
2018, https://www.cnet.com/news/google-explains-gmail-privacy-after-controversy/.

35 Allan Smith, "Facebook, Amazon, and Google Just Spent Record Amounts of Cash on
Lobbying Washington, DC," *Business Insider*, July 23, 2018, https://www.businessin-
sider.com/amazon-facebook-google-spend-record-amounts-lobbying-2018-7.

36 Mims, "Who Has More."

37 David Goodfriend, "Stand With the Alt-Right Against Google? No, Thanks," *Minn-
Post*, June 26, 2018, https://www.minnpost.com/community-voices/2018/06/
stand-alt-right-against-google-no-thanks.

38 "Sinclair Broadcast Group's Revenue From 2009 to 2018," Statista.com, https://www.
statista.com/statistics/316998/sinclair-broadcasting-annual-revenue/.

39 "Google's Revenue Worldwide From 2002 to 2018," Statista.com, https://www.statista.
com/statistics/266206/googles-annual-global-revenue/.

40 Matt Hunter and Anita Balakrishnan, "Apple's Cash Pile Hits $285.1 Billion, a
Record," *CNBC.com*, Feb. 1, 2018, https://www.cnbc.com/2018/02/01/apple-earn-
ings-q1-2018-how-much-money-does-apple-have.html; List of Countries by GDP. In
Wikipedia, https://en.wikipedia.org/wiki/List_of_countries_by_GDP_(nominal).

41 Eric Lichtblau and Katie Benner, "Apple Fights Order to Unlock San Bernardino
Gunman's iPhone," *The New York Times*, Feb. 17, 2016, https://www.nytimes.
com/2016/02/18/technology/apple-timothy-cook-fbi-san-bernardino.html.

42 Juli Clover, "Apple CEO Tim Cook Talks Immigration," *Mac Rumors*, June 25, 2018,
https://www.macrumors.com/2018/06/25/cook-social-issues-interview-fortune/.

43 Galloway, *The Four.*

44 Lisa W. Foderaro, "Private Moment Made Public, Then a Fatal Jump," *The New York
Times*, Sept. 29, 2010, https://www.nytimes.com/2010/09/30/nyregion/30suicide.html.

45 "U.S. Teenagers Charged Over Suicide of Irish 'New Girl' Targeted in 'Relentless'
School Bullying Campaign," *The Daily Mail*, March 31, 2010, http://www.dailymail.
co.uk/news/article-1262487/Phoebe-Prince-9-US-teenagers-charged-suicide-death-
Irish-new-girl.html.

46 Gwenn Schurgin O'Keeffe, Kathleen Clarke-Pearson, "The Impact of Social Media on
Children, Adolescents, and Families," 127: 4, *AAP News & Journals Gateway: Pediatrics*, April 2011, https://pediatrics.aappublications.org/content/127/4/800.

47 "Man Live Streams Suicide on Facebook," *The Hindu*, Aug. 1, 2018, https://www.thehindu.
com/news/cities/Delhi/man-live-streams-suicide-on-facebook/article24567463.ece.

48 Izzy Kalman, "Who Really Killed 11 Year Old Jaheem
Herrera?" *Psychology Today*, Aug. 27, 2009,https://www.psycholo-
gytoday.com/us/blog/the-psychological-solution-bullying/200908/
who-really-killed-11-year-old-jaheem-herrera.

49 Betsy Morris, "The New Tech Avengers," *The Wall Street Journal*, June 29, 2018, https://www.wsj.com/articles/the-new-tech-avengers-1530285064.

50 "Howard Gardner," Harvard Graduate School of Education website, https://www.gse.harvard.edu/faculty/howard-gardner

51 James P. Steyer, *Talking Back to Facebook* (New York: Simon & Shuster, 2012).

52 Amy O'Leary, "So How Do We Talk About This? When Children See Internet Pornography," *The New York Times*, May 9, 2012, https://nytimes.com/2012/05/10/garden/when-children-see-internet-pornography.html.

53 Jonathan Vanian, "How Data Privacy Blunders and Conspiracy Theories Helped Fuel the 'Techlash'," *Fortune*, July 17, 2017, http://fortune.com/2018/07/17/techlash-brainstorm-privacy-conspiracy/.

54 Ibid.

55 Morris, "The New Tech Avengers."

56 Rachel Botsman, *Who Can You Trust? How Technology Brought Us Together and Why It Might Drive Us Apart* (New York: Public Affairs, 2017).

57 Betsy Morris, "Most Teens Prefer to Chat Online, Rather Than in Person," *The Wall Street Journal*, Sept. 10, 2018, https://www.wsj.com/articles/most-teens-prefer-to-chat-online-than-in-person-survey-finds-1536597971.

58 Steyer, *Talking Back.*

59 Hayley Tsukuyama, "Teens Spend Nearly Nine Hours Every Day Consuming Media," Nov. 3, 2015, *The Washington Post*, https://www.washingtonpost.com/news/the-switch/wp/2015/11/03/teens-spend-nearly-nine-hours-every-day-consuming-media.

60 Jacqueline Howard, "ADHD Study Links Teens' Symptoms With Digital Media Use," CNN.com, July 17, 2018, https://www.cnn.com/2018/07/17/health/adhd-symptoms-digital-media-study/index.html.

61 Steyer, *Talking Back.*

62 "David Meyer," University of Michigan. website, https://lsa.umich.edu/psych/people/faculty/demeyer.html.

63 Daniela Hernandez and Betsy Morris, "Frequent Technology Use Linked to ADHD Symptoms in Teens, Study Finds," *The Wall Street Journal*, July 27, 2018, https://www.wsj.com/articles/frequent-technology-use-linked-to-adhd-symptoms-in-teens-study-1531839628.

64 Jean M. Twenge, *iGen: Why Today's Super-Connected Kids Are Growing Up Less Rebellious, More Tolerant, Less Happy—and Completely Unprepared for Adulthood—and What That Means for the Rest of Us* (New York: Simon & Schuster, 2017).

65 Hernandez and Morris, "ADHD Symptoms in Teens."

66 Nicholas Carr, *The Shallows: What the Internet Is Doing to Our Brains* (New York: W.W. Norton, 2011).

67 LaVonne Neff, "How Our Minds Have Changed," *Christian Century*, Sept. 22, 2010, https://www.christiancentury.org/reviews/2010-09/how-our-minds-have-changed.

68 Chris Weller, "Bill Gates and Steve Jobs Raised Their Kids Tech-Free—and It Should've Been a Red Flag," *Insider*, Oct. 24, 2017, https://www.thisisinsider.com/screen-time-limits-bill-gates-steve-jobs-red-flag-2017-10.

69 Mike Allen, "Sean Parker Unloads on Facebook: 'God Only Knows What It's Doing to Our Children's Brains,'" *Axios*, Nov. 9, 2017, https://www.axios.com/sean-parker-unloads-on-facebook-2508036343.html.

70 Jennings Brown, "Former Facebook Exec: 'You Don't Realize It But You Are Being Programmed,'" *Gizmodo*, Dec. 11, 2017, https://gizmodo.com/former-facebook-exec-you-don-t-realize-it-but-you-are-1821181133.

71 Vanian, "Techlash."

72 Alexandra Ma, "A Sad Number of Americans Sleep With Their Smartphone in Their Hand," *HuffPost*, June 29, 2015, https://www.huffingtonpost.com/2015/06/29/smart-phone-behavior-2015_n_7690448.html.

73 Amanda Lenhart, Rich Ling, Scott Campbell, and Kristen Purcell, "Teens and Mobile Phones," report, Pew Research Center, April 20, 2010, http://www.pewinternet.org/2010/04/20/teens-and-mobile-phones/.

74 Alina Bradford, "How Blue LEDs Affect Sleep," *Live Science*, Feb. 26, 2016, https://www.livescience.com/53874-blue-light-sleep.html.

75 Matt Richtel, "Drivers and Legislators Dismiss Cellphone Risks," *The New York Times*, July 18, 2009, https://www.nytimes.com/2009/07/19/technology/19distracted.html.

76 "Texting and Driving Accident Statistics," Edgar Snyder & Associates law firm website, https://www.edgarsnyder.com/car-accident/cause-of-accident/cell-phone/cell-phone-statistics.html.

77 Ibid.

78 Christian Nordqvist, "One Million People Commit Suicide Each Year," *Medical News Today*, Sept. 10, 2011, https://www.medicalnewstoday.com/articles/234219.php.

79 Michal Lev-Ram, "Facebook's Fix-It Team," *Fortune*, May 22 2018, http://fortune.com/longform/facebook-fix-it-team-fortune-500/.

80 Schmidt, Eric. In WikiQuote, https://en.wikiquote.org/wiki/Eric_Schmidt.

81 Lee Humphreys, *The Qualified Self: Social Media and the Accounting of Everyday Life* (Cambridge, Mass.: MIT Press, 2018).

82 "CR Survey: 7.5 Million Facebook Users Are Under the Age of 13, Violating the Site's Terms," *Consumer Reports*, May 10, 2011, https://www.consumerreports.org/media-room/press-releases/2011/05/cr-survey-75-million-facebook-users-are-under-the-age-of-13-violating-the-sites-terms-/.

83 Brian Y. Park, Gary Wilson, Jonathan Berger, Matthew Christman, Bryn Reina, Frank Bishop, Warren P. Klam, and Andrew P. Doan, "Is Internet Pornography Causing Sexual Dysfunctions? A Review With Clinical Reports," *Behavioral Sciences* 6, No. 3 (Sept. 2016), 17.

84 Philip Perry, "How Internet Porn Is Changing the Way Men and Women Are Having Sex," *Big Think*, Oct. 21, 2017, http://bigthink.com/philip-perry/how-internet-porn-is-changing-how-men-and-women-are-having-sex.

85 Katherine Wu, "Love, Actually: The Science Behind Lust, Attraction and Companionship," Harvard University Graduate School of Arts and Sciences Blog, Feb. 14, 2017, http://sitn.hms.harvard.edu/flash/2017/love-actually-science-behind-lust-attraction-companionship/.

86 O'Leary, "So How Do We Talk About This?"

87 Henry Timms and Jeremy Heimans, "#DeleteFacebook Is Just the Beginning. Here's the Movement We Could See Next," *Fortune*, April 16, 2018, http://fortune.com/2018/04/16/delete-facebook-data-privacy-movement/.

88 Kranzberg, Melvin. In Wikipedia, https://en.wikipedia.org/wiki/Melvin_Kranzberg.

STRIKE 2: STASI TACTICS

1   Kashmir Hill, "When a Stranger Decides to Destroy Your Life," *Gizmodo*, July 26, 2018, https://gizmodo.com/when-a-stranger-decides-to-destroy-your-life-1827546385

2   Richard Chirgwin, "Tim? Larry? We Need to Talk About Smartphones and Privacy," *The Register*, July 12, 2018, https://www.theregister.co.uk/2018/07/12/us_lawmakers_demand_privacy_answers_from_apple_and_alphabet.

3   Kent Ninomiya, "Invasion of Privacy on Social Media," LinkedIn.com, May 3, 2016, https://www.linkedin.com/pulse/invasion-privacy-social-media-kent-ninomiya-1/.

4   John Leonard, "Facebook Privacy Loophole Allowed Personal Data of 'Closed' Group Members to Be Downloaded," *Computing*, July 13, 2018, https://www.computing.co.uk/ctg/news/3035843/facebook-privacy-loophole-allowed-personal-data-of-closed-group-members-to-be-downloaded.

5   David Pearson, "Facebook Needed Third-Party Apps to Grow. Now It's Left With a Privacy Crisis," *Los Angeles Times*, March 20, 2018, https://www.latimes.com/business/technology/la-fi-tn-facebook-third-parties-20180320-story.html.

6   Dave Eggers, *The Circle* (New York: Random House, (2013).

7   Julia Angwin, *Dragnet Nation* (New York: Henry Holt, 2014).

8   Cameron F. Kelly, "Why Protecting Privacy Is a Losing Game Today—and How to Change the Game," Brookings Institution report, July 12, 2018, https://www.brookings.edu/research/why-protecting-privacy-is-a-losing-game-today-and-how-to-change-the-game/.

9   Nicholas Confessore, Gabriel J.X. Dance, Richard Harris, and Mark Hansen, "The Follower Factory," *The New York Times*, Jan. 27, 2018, https://www.nytimes.com/interactive/2018/01/27/technology/social-media-bots.html.

10 Ibid.

11 Steven Melendez, "Cybersecurity Pros Are Limiting Their Personal Use of Facebook, Survey Says," *Fast Company*, June 26, 2018, https://www.fastcompany.com/40589927/cybersecurity-pros-are-limiting-their-personal-use-of-facebook-survey-says.

12 Luke Larson, "Facebook Wants to Own Your Face. Here's Why That's a Privacy Disaster," *Digital Trends*, July 13, 2018, https://www.digitaltrends.com/computing/facebook-facial-recognition-privacy/.

13 "Google's New Facial Recognition Patent Wants to Stalk Your Social Media," *CB Insights*, July 31, 2018, https://www.cbinsights.com/research/google-facial-recognition-social-media-patent.

14 Christopher Mims, "Amazon's Face-Scanning Surveillance Software Contrasts With Its Privacy Stance," *The Wall Street Journal*, June 21, 2018, https://www.wsj.com/articles/the-privacy-paradox-face-recognition-is-techs-next-moral-dilemma-1529596801.

15 Scott Thurm, "Microsoft Calls for Federal Regulation of Facial Recognition," *Wired*, July 13, 2018, https://www.wired.com/story/microsoft-calls-for-federal-regulation-of-facial-recognition/?CNDID=48193078&mbid=nl_071318_daily_list1_p1.

16 Anthony Cuthbertson, "Facebook Patent Predicts When You'll Die," *The Independent*, June 26, 2018, https://www.independent.co.uk/life-style/gadgets-and-tech/

news/facebook-patent-predict-die-death-prediction-algorithm-personal-data-priva-cy-a8417771.html.

17   Tegan Jones, "Facebook Says It Will Never Use Secret Recording Technology It Filed a Patent For," *Gizmodo*, June 28, 2018,https://www.gizmodo.com.au/2018/06/facebook-says-it-will-never-use-secret-recording-technology-it-filed-a-patent-for/.

18   Joel Hruska, "Facebook Files Patent for Exactly the Kind of Spying It Claims It Doesn't Do," *Extreme Tech*, June 29, 2018, https://www.extremetech.com/computing/272630-facebook-files-patent-spying.

19   Ibid.

20   Mims, "Amazon's Face-Scanning."

21   Angwin, *Dragnet Nation.*

22   Letter the eight groups sent to Federal Trade Commission, June 27, 2018, http://thepublicvoice.org/wp-content/uploads/2018/06/FTC-letter-Deceived-by-Design.pdf.

23   Forbrukerrådet (Norway Council of Consumers), "Deceived by Design," June 27, 2018,   http://www.consumerwatchdog.org/sites/default/files/2018-06/2018-06-25%20Deceived%20by%20design%20-%20Final.pdf.

24   Lee Fang, "Google and Facebook Are Quietly Fighting California's Privacy Rights Initiative, Emails Reveal," *The Intercept*, June 27, 2018, https://theintercept.com/2018/06/26/google-and-facebook-are-quietly-fighting-californias-privacy-rights-initiative-emails-reveal/.

25   Ibid.

26   Estimates of automated traffic on the internet vary widely, since it is a distributed system. But 36 percent is often cited based on an IAB study: http://www.digitalmarketing-glossary.com/What-is-Bot-traffic-definition//dead link//

27   Confessore et al., "The Follower Factory."

28   Kent Ninomiya, "Invasion of Privacy on Social Media," LinkedIn.com, May 3, 2016, https://www.linkedin.com/pulse/invasion-privacy-social-media-kent-ninomiya-1/.

29   David Kravets, "FTC Dings Google $22.5M in Safari Cookie Flap," *Wired*, Aug. 9, 2012, https://www.wired.com/2012/08/ftc-google-cookie/.

30   "Google's Revenue."

31   Greg Ip, "The Antitrust Case Against Facebook, Google and Amazon," *The Wall Street Journal*, Jan. 16, 2018, https://www.wsj.com/articles/the-antitrust-case-against-facebook-google-amazon-and-apple-1516121561.

32   Charles Duhigg, "The Case Against Google," *The New York Times Magazine*, Feb. 20, 2018, https://www.nytimes.com/2018/02/20/magazine/the-case-against-google.html.

33   David Dayen, "The Android Administration," *The Intercept*, April 22, 2016, https://theintercept.com/2016/04/22/googles-remarkably-close-relationship-with-the-obama-white-house-in-two-charts/.

34   Steven Levy, *In the Plex: How Google Thinks, Works, and Shapes Our Lives* (New York: Simon & Schuster, 2011).

35   Adam J. White, a Hoover Institution fellow, is also the director of the Center for the Study of the Administrative State at George Mason University's Antonin Scalia Law School. His complete article "Google.gov" offers a number of valuable insights into Google's past and trajectory, as well, and is recommended in its entirety. Find it at: https://www.thenewatlantis.com/publications/googlegov

Or, Adam J. White, "Google.gov," *The New Atlantis* No. 55 (Spring 2018), p. 3-34.

36   Patrick Holland, "Obama Warns Facebook and Google to Check Themselves," *CNet*, Feb. 27, 2018, https://www.cnet.com/news/barack-obama-warns-facebook-and-google-during-mit-speech/.

37   White, "Google.gov."

38   "Google Search Statistics," accessed April 11, 2019, Internet Live Stats, http://www.internetlivestats.com/google-search-statistics/; "Worldwide Net Mobile Advertising Revenues of Google From 2014 to 2018," Statista.com, https://www.statista.com/statistics/539477/google-mobile-ad-revenues-worldwide/.

39   Daniel Liberto, "Facebook, Google Digital Ad Market Share Drops as Amazon Climbs," Investopedia, March 20, 2018, https://www.investopedia.com/news/facebook-google-digital-ad-market-share-drops-amazon-climbs/.

40   Benjamin Edelman, "Hard-Coding Bias in Google 'Algorithmic' Search Results," BenEdelman.org, Nov. 15, 2010, http://www.benedelman.org/hardcoding/.

41   Kashmir Hill, "Google Gets Grief From Senators, Mimes, and an Ice Cream Truck," *Forbes*, Sept. 21, 2011, https://www.forbes.com/sites/kashmirhill/2011/09/21/google-gets-grief-from-senators-mimes-and-an-ice-cream-truck/#8abf8fa53956.

42   "Barack Obama's Speech at Google," *Blogoscope*, Nov. 19, 2007, http://blogoscoped.com/archive/2007-11-19-n10.html.

43   Claudia Geib, "Obama: Social Media Is 'Shaping Our Culture in Powerful Ways'," *Futurism*, Feb. 28, 2018, https://futurism.com/obama-social-media-video/.

44   "What Is a Trust Indicator?" Frequently Asked Questions, *The Trust Project*, accessed April 11, 2019, https://thetrustproject.org/faq/#indicator.

45   "If every person on earth was to be given an equal portion of inhabitable land, how much land would each person get?" Quora.com, accessed April 11, 2019, https://www.quora.com/If-every-person-on-earth-was-to-be-given-an-equal-portion-of-inhabitable-land-how-much-land-would-each-person-get AND http://www.rain-tree.com/facts.htm#.W3Xicy3MyL5; "How many countries does Amazon operate inside of, and what is the revenue ranked per country?" Quora.com, accessed April 11, 2019, https://www.quora.com/How-many-countries-does-Amazon-operate-inside-of-and-what-is-the-revenue-ranked-per-country.

46   *The Economist* reports Brazil earns $10 billion a year from soy, coffee, sugar, lumber and other forest products. Statistics for other countries are not well known, so total Amazon rainforest commercial output is estimated based on land-size equivalents in the six countries the Amazon encompasses. "The Economy Booms, the Trees Vanish," *The Economist*, May 19, 2005, https://www.economist.com/news/2005/05/19/the-economy-booms-the-trees-vanish.

47   Total return to shareholders assumes the 2007-2017 annual rate.

48   Erica Pandey, "The Percentage of U.S. Households With Guns Is Falling," *Axios*, Oct. 3, 2017, https://www.axios.com/the-percentage-of-us-households-with-guns-is-falling-1513305943-490b2051-3056-4020-ac55-0d091641d80f.html; Shep Hyken, "Sixty-Four Percent of U.S. Households Have Amazon Prime," *Fortune*, June 17, 2017, https://www.forbes.com/sites/shephyken/2017/06/17/sixty-four-percent-of-u-s-households-have-amazon-prime/#4ff545164586.

49   Krystina Gustafson, "Amazon Captured More Than Half of All Online Sales Growth Last Year, New Data Shows," CNBC.com, Feb. 1, 2017, https://www.cnbc.com/2017/02/01/amazon-captured-more-than-half-of-all-online-sales-growth-last-year.html.

50   Paul Grey, "How Many Products Does Amazon Sell?" ExportX.com, Dec. 11, 2015, https://export-x.com/2015/12/11/how-many-products-does-amazon-sell-2015/; "Rainforest Facts," Raintree.com, accessed April 11, 2019, http://www.rain-tree.com/facts.htm#.W3Xggi3MyL4.

51   "Amazon Revenue 2006-2018," MacroTrends, accessed April 11, 2019, https://www.macrotrends.net/stocks/charts/AMZN/amazon/revenue.

52   McKinsey Global Institute, "Jobs Lost, Jobs Gained: Workforce Transitions in a Time of Automation," December 2017.

53   Galloway, *The Four*.

54   Ry Crist, "Alexa Sent Private Audio to a Random Contact, Portland Family Says," *CNet*, May 24, 2018, https://www.cnet.com/news/alexa-sent-private-audio-to-a-random-contact-portland-family-says/.

55   Tara Seals, "'Voice-Squatting' Turns Alexa, Google Home into Silent Spies," *Threatpost*, May 17, 2018, https://threatpost.com/voice-squatting-turns-alexa-google-home-into-silent-spies/132068/.

56   Galloway, *The Four*.

57   Beth Kowitt, "How Amazon Is Using Whole Foods in a Bid for Total Retail Domination," *Fortune*, May 21, 2018, http://fortune.com/longform/amazon-groceries-fortune-500/.

58   Dan O'Shea, "Survey: 55% of Shoppers Turn to Amazon to Begin Product Search," *Retail Dive*, Sept. 27, 2016, https://www.retaildive.com/news/survey-55-of-shoppers-turn-to-amazon-to-begin-product-search/427118/.

59   Katie Richards, "HP, Ikea and Audi Are Among the Top 100 Brands That Consumers Trust Most," *Ad Week*, Oct. 17, 2017, https://www.adweek.com/brand-marketing/hp-ikea-and-audi-are-among-the-top-100-brands-that-consumers-trust-most-in-the-u-s/.

60   Lauren Thomas and Courtney Regan, "Watch Out, Retailers. This Is Just How Big Amazon Is Becoming," *CNBC.com*, July 13, 2018, https://www.cnbc.com/2018/07/12/amazon-to-take-almost-50-percent-of-us-e-commerce-market-by-years-end.html.

61   Pat Ahern, "25 Mind-Bottling SEO Stats for 2019 (and Beyond)," *Junto*, Last Updated April 8, 2019, https://junto.digital/blog/seo-stats-2017/.

62   Kowitt, "Total Retail Domination." AND Melissa Anders, "Retail Industry Expects More Sales Growth In 2018," *Forbes*, Feb. 8, 2018, https://www.forbes.com/sites/melissaanders/2018/02/08/retail-industry-poised-for-more-sales-growth-in-2018/#5cc3fb3a4fa7.

63   Gerard du Toit and Aaron Cheris, "Banking's Amazon Moment," Bain & Company website, March 5, 2018, https://www.bain.com/insights/bankings-amazon-moment/.

64   "U.S. Drug Store/Pharmacy Market," Statista.com, accessed April 11, 2019, https://www.statista.com/topics/1412/drug-store-pharmacy-market-in-the-us/.

65   "Event Tickets," Statista.com, accessed April 11, 2019, https://www.statista.com/outlook/264/109/event-tickets/united-states; "Online Event Ticketing Market Size Worth $67.99 Billion By 2025," Grand View Research, May 2018, https://www.grandviewresearch.com/press-release/global-online-event-ticketing-market.

66　Sam Dangremond, "Jeff Bezos Is Renovating the Biggest House in Washington, D.C.," *Town & Country,* April 24, 2018 (updated April 4, 2019), https://www.townand-countrymag.com/leisure/real-estate/news/a9234/jeff-bezos-house-washington-dc/.

STRIKE 3: THOUGHT MANIPULATION

1　Katherine Losse, *The Boy Kings: A Journey Into the Heart of the Social Network* (New York: Free Press/Simon & Schuster 2012).

2　Natasha Lomas, "A Brief History of Facebook's Privacy Hostility Ahead of Zuckerberg's Testimony," *Tech Crunch,* accessed April 11, 2019, https://techcrunch.com/2018/04/10/a-brief-history-of-facebooks-privacy-hostility-ahead-of-zuckerbergs-testimony/.

3　Anita Balakrishnan, Sara Salinas and Matt Hunter, "Mark Zuckerberg Has Been Talking About Privacy for 15 Years—Here's Almost Everything He's Said," *CNBC.com,* April 9, 2018, https://www.cnbc.com/2018/03/21/facebook-ceo-mark-zuckerbergs-statements-on-privacy-2003-2018.html.

4　"In the Matter of Facebook, Inc., a Corporation," Federal Trade Commission consent order, accessed April 11, 2019, https://www.ftc.gov/sites/default/files/documents/cases/2011/11/111129facebookagree.pdf Wired story on potential fines: Nitasha Tiku, "Why Facebook's 2011 Promises Haven't Protected Users," *Wired,* April 11, 2018, https://www.wired.com/story/why-facebooks-2011-promises-havent-protected-users.

5　Rachel Botsman, *Who Can You Trust? How Technology Brought Us Together and Why It Might Drive Us Apart* (New York: Public Affairs, 2017).

6　Ibid.

7　Alexis C. Madrigal, "What Facebook Did to American Democracy," *The Atlantic,* Oct. 12, 2017, https://www.theatlantic.com/technology/archive/2017/10/what-facebook-did/542502/.

8　Pomerantsev, Peter. In Wikipedia, https://en.wikipedia.org/wiki/Peter_Pomerantsev.

9　Peter Pomerantsev, *Nothing Is True and Everything Is Possible: The Surreal Heart of the New Russia* (Kindle Edition: Public Edition, 2014).

10　Curt Levey, "Why Should We Fear Russian Political Ads?" *The Wall Street Journal,* Aug. 16, 2018,https://www.wsj.com/articles/why-should-we-fear-russian-political-ads-1534460641.

11　Madrigal, "What Facebook Did."

12　Liam Tung, "Facebook: We're Adding Information Warfare to Our Fight Against Malware, Fraud," *ZDNet,* April 28, 2017, https://www.zdnet.com/article/facebook-were-adding-information-warfare-to-our-fight-against-malware-fraud/.

13　Sean J. Miller, "Digital Ad Spending Tops Estimates," *Campaigns & Elections,* January 4, 2017, https://www.campaignsandelections.com/campaign-insider/digital-ad-spending-tops-estimates.

14　Facebook, "Facebook Brussels," May 23, 2018, https://www.facebook.com/facebookbrussels/posts/follow-up-questions-from-epduring-his-meeting-with-the-conference-of-presidents-/1769490786430772/.

15　George Upper and Shaun Hair, "Confirmed: Facebook's Recent Algorithm Change Is Crushing Conservative Sites, Boosting Liberals," *The Western Journal,* March 13, 2018, https://www.westernjournal.com/confirmed-facebooks-recent-algorithm-change-is-crushing-conservative-voices-boosting-liberals/.

16 John Herrman, "Inside Facebook's (Totally Insane, Unintentionally Gigantic, Hyperpartisan) Political-Media Machine," *The New York Times Magazine*, Aug. 27, 2016, https://www.nytimes.com/2016/08/28/magazine/inside-facebooks-totally-insane-unintentionally-gigantic-hyperpartisan-political-media-machine.html.

17 Madrigal, "What Facebook Did."

18 Nicholas Thompson and Fred Vogelstein, "Inside the Two Years That Shook Facebook—and the World," *Wired*, Feb. 12, 2018, https://www.wired.com/story/inside-facebook-mark-zuckerberg-2-years-of-hell/.

19 Lindsey Bever, "Mark Zuckerberg Pledges 'to Do the Job He Already Has,' Basically," *The Washington Post*, Jan. 4, 2018, https://www.washingtonpost.com/news/the-switch/wp/2018/01/04/mark-zuckerberg-pledges-to-do-the-job-he-already-has-basically; Erin Kelly, "Republicans Press Social Media Giants on Anti-Conservative 'Bias' That Dems Call 'Nonsense,'" *USA Today*, July 17, 2018, https://www.usatoday.com/story/news/politics/2018/07/17/gop-lawmakers-press-facebook-twitter-anti-conservative-bias/792555002; Garett Sloane, "'Milkshakes Against the Republican Party' Disappears After Congress Pressures Facebook," *Ad Age*, republished by Marcus Myles Media, July 18, 2018, https://marcusmylesmedia.com/media-buying-boston/milkshakes-against-the-republican-party-disappears-after-congress-pressures-facebook/; Benjamin Mullin and Deepa Seetharaman, "Lawmakers Question Tech Firms, Publishers on How They Combat Fake News," *The Wall Street Journal*, July 18, 2018, https://www.wsj.com/articles/lawmakers-question-tech-firms-publishers-on-how-they-combat-fake-news-1531867010.

20 Christopher Mims, "What Mark Zuckerberg Didn't Say About What Facebook Knows About You," *The Wall Street Journal*, April 14, 2018, https://www.wsj.com/articles/what-mark-zuckerberg-didnt-say-about-what-facebook-knows-about-you-1523726008.

21 Ibid.

22 Mark Zuckerberg post on Facebook, Jan. 4, 2018, https://www.facebook.com/zuck/posts/10104380170714571.

23 Bloomberg Government, "Transcript of Mark Zuckerberg's Senate hearing," *The Washington Post*, April 10, 2018, https://www.washingtonpost.com/news/the-switch/wp/2018/04/10/transcript-of-mark-zuckerbergs-senate-hearing/?noredirect=on&utm_term=.8450e25735f0.

24 Kurt Wagner, "I Designed Facebook, 'So if Someone's Going to Be Fired for This, It Should Be Me,'" *Recode*, July 18, 2018, https://www.recode.net/2018/7/18/17585918/facebook-cambridge-analytica-mark-zuckerberg-responsibility-fired.

25 "Facing Facts," video on Facebook, accessed April 11, 2019, https://newsroom.fb.com/news/2018/05/inside-feed-facing-facts/.

26 Madrigal, "What Facebook Did."

27 Edelman, "2018 Edelman Trust Barometer Reveals Record-Breaking Drop in Trust in the U.S.," January 22, 2018, https://www.edelman.com/news-awards/2018-edelman-trust-barometer-reveals-record-breaking-drop-trust-in-the-us.

28 Craig Timberg, Elizabeth Dwoskin, Matt Zapotosky and Devlin Barrett, "Facebook's Disclosures Under Scrutiny as Federal Agencies Join Probe of Tech Giant's Role in Sharing Data With Cambridge Analytica," *The Washington Post*, July 2, 2018, https://www.washingtonpost.com/technology/2018/07/02/federal-investigators-

broaden-focus-facebooks-role-sharing-data-with-cambridge-analytica-examining-statements-tech-giant.

29   "About the California Consumer Privacy Act," Californians for Consumer Privacy, accessed April 11, 2019, https://www.caprivacy.org/about.

30   Will Oremus, "The Great Facebook Crash," *Slate*, June 27, 2018, https://slate.com/technology/2018/06/facebooks-retreat-from-the-news-has-painful-for-publishers-including-slate.html.

31   Calvin Freiburger, "Huh? Left-Wing Media Watchdog Accuses Facebook of 'Caving' to Conservatives," *LifeSite*, June 26, 2018,
https://www.lifesitenews.com/news/media-matters-accuses-facebook-of-caving-to-conservatives-in-cnn-piece.

32   Joe Schoffstall, "Read the Confidential David Brock Memo Outlining Plans to Attack Trump," *Washington Free Beacon*, Jan. 26, 2017, https://freebeacon.com/politics/david-brock-memo-attack-trump/.

33   Jim Holt, "Top Far Left Organizations Bragged About Working with Facebook and Twitter to Censor and Eliminate Conservative Content," *The Gateway Pundit*, Aug. 20, 2018, https://www.thegatewaypundit.com/2018/08/top-democrat-activist-organizations-admitted-to-working-with-facebook-and-twitter-to-eliminate-conservative-content/.

34   Aaron Bandler, "11 of the Most Incredible Things From David Brock's Uncovered Playbook," *The Daily Wire*, Aug. 7, 2017, https://www.dailywire.com/news/19445/11-most-incredible-things-media-matters-uncovered-aaron-bandler.

35   Holt, "Top Far Left."

36   "Censored! How Online Media Companies Are Suppressing Conservative Speech," Media Research Center, accessed April 11, 2019, https://cdn.mrc.org/static/censored/mrc-censorship-report.pdf.

37   JD Heyes, "Facebook Slammed for Once Again Censoring Content From Pro-America Country Music Group's Song, 'I Stand for the Flag,'" Censorship News, July 5, 2018, http://censorship.news/2018-07-05-facebook-slammed-censoring-pro-america-country-music-groups-song-i-stand-for-the-flag.html.

38   Donald Trump (@realDonaldTrump). Twitter post, Aug. 18, 2018, https://twitter.com/realDonaldTrump/status/1030777074959757313.

39   Trump, Twitter post, Aug. 18, 2018, https://twitter.com/realdonaldtrump/status/1030779412973846529.

40   Trump, Twitter post, Aug. 18, 2018, https://twitter.com/realDonaldTrump/status/1030777074959757313.

41   Trump, Twitter post, Aug. 18, 2018, https://twitter.com/realdonaldtrump/status/1030781399920455681.

42   Michael W. Chapman, "Alveda King Says Facebook Pulled Ads for Pro-Life 'Roe v. Wade' Movie," *CNS News*, Jan. 17, 2018, https://www.cnsnews.com/news/article/michael-w-chapman/dr-alveda-king-facebook-has-pulled-down-our-ads-abortion-its-very.

43   "Nick Loeb on Newsmax TV discussing ROE v. WADE the Movie," YouTube post, Jan. 15, 2018, https://www.bing.com/videos/search?q=nick+loeb&view=detail&mid=958DCD35CA3A37963756958DCD35CA3A37963756&FORM=VIRE.

44   "Matt Margolis," MattMargolis.com, accessed April 11, 2019, https://mattmargolis.com/about/.

45   Jayson Veley, "Continuing Its Tirade of Political Censorship, Facebook Just BANNED the Author of a New Book That criticizes Obama," *Censorship News*, Jan. 15, 2018, http://www.censorship.news/2018-01-15-political-censorship-facebook-just-banned-the-author-of-a-new-book-that-criticizes-obama.html.

46   Megan Fox, "Facebook Bans Bestselling Author Over 'The Scandalous Presidency of Barack Obama'," *PJ Media*, Jan. 11, 2018, https://pjmedia.com/trending/facebook-bans-bestselling-author-ad-scandalous-presidency-barack-obama/.

47   Michael Nunez, "Former Facebook Workers: We Routinely Suppressed Conservative News," *Gizmodo*, May 9, 2016, https://gizmodo.com/former-facebook-workers-we-routinely-suppressed-conser-1775461006.

48   Alex Hern, "Facebook Denies Censoring Conservative Stories From Trending Topics," *The Guardian*, May 10, 2016, https://www.theguardian.com/technology/2016/may/10/facebook-denies-censoring-conservative-trending-news.

49   "CNN Purchases Industrial-Sized Washing Machine to Spin News Before Publication," *The Babylon Bee*, March 1, 2018, https://babylonbee.com/news/cnn-purchases-industrial-sized-washing-machine-spin-news-publication/.

50   Joel B. Pollack, "Facebook Allows 'Fake News' Photo to Raise $18 Million for Border Cause," *Breitbart*, June 21, 2018, https://www.breitbart.com/border/2018/06/21/facebook-allows-fake-news-photograph-to-raise-18-million-for-border-cause.

51   Lilly Price, "Crying Immigrant Girl on Time Magazine Cover Was Never Separated From Her Mom, Father Says," *USA Today*, June 22, 2018, https://www.usatoday.com/story/news/nation/2018/06/22/girl-time-magazine-cover-never-separated-mom-father-says/725499002/.

52   Casey Ryan, "Facebook Censors His Conservative Posts, Retired Accountant Contends," *The Daily Signal*, Oct. 23, 2017, http://dailysignal.com/2017/10/23/facebook-censors-his-conservative-posts-retired-accountant-contends/.

53   Mary Rezac, "Two Dozen Catholic Pages Blocked From Facebook Without Explanation," *Catholic News Agency*, July 18, 2017, https://www.catholicnewsagency.com/news/several-catholic-pages-blocked-from-facebook-without-explanation-59785.

54   Ibid.

55   Rebecca Tushnet, "Power Without Responsibility: Intermediaries and the First Amendment," *George Washington Law Review* 76, No. 4 (June 2008), 986-1016. http://www.gwlr.org/wp-content/uploads/2012/08/76-4-Tushnet.pdf

56   "Marsh v. Alabama," Cornell Law School Legal Information Institute, accessed April 11, 2019, https://www.law.cornell.edu/supremecourt/text/326/501.

57   Peter Hasson, "Exclusive: Facebook, Amazon, Google and Twitter All Work with Left-Wing SPLC," *The Daily Caller*, June 6, 2018, https://dailycaller.com/2018/06/06/splc-partner-google-facebook-amazon/.

58   Valerie Richardson, "Mainstream Conservative Groups Alarmed to Be Found on 'Hate Map'," *The Washington Times*, Aug. 17, 2017, https://www.washingtontimes.com/news/2017/aug/17/hate-group-map-lists-mainstream-conservative-organ/.

59   Freiburger, "Huh?"

60   Among the cases Alliance Defending Freedom was honored to argue before the U.S. Supreme Court, the Masterpiece Cakeshop v. Colorado Civil Rights Commission and National Institute for Family and Life Advocates v. Becerra both were clarion calls for free expression in a free nation.

61   Michael Farris, "Southern Poverty Law Center 'Hate' Labels Deserve a Vigorous Response," *National Review*, Aug. 17, 2018, https://www.nationalreview.com/2018/08/southern-poverty-law-center-hate-label-deserves-vigorous-response.

62   Ibid.

63   Calvin Freiburger, "60 Groups Consider Suing Southern Poverty Law Center After It Pays $3.4 M in Defamation Settlement," *Life Site*, June 20, 2018, https://www.lifesitenews.com/news/3.4-million-defamation-win-inspires-60-groups-to-consider-suing-anti-christ.

64   Adam Liptak, "How Conservatives Weaponized the First Amendment," *The New York Times*, June 30, 2018, https://www.nytimes.com/2018/06/30/us/politics/first-amendment-conservatives-supreme-court.html.

65   Carolanne Mangles, "Search Engine Statistics 2018," *Smart Insights*, Jan. 30, 2018, https://www.smartinsights.com/search-engine-marketing/search-engine-statistics/.

66   Andy Long, "Trump Left Out of Google Search for Presidential Candidates," *NBC4i.com*, July 27, 2016, https://www.nbc4i.com/news/politics/trump-left-out-of-google-search-for-presidential-candidates/1114620384.

67   Steve Goldstein and Jeremy C. Owens, "Google Searches for Hillary Clinton Yield Favorable Autocomplete Results, Report Shows," *MarketWatch*, June 9, 2016, https://www.marketwatch.com/story/google-searches-for-hillary-clinton-yield-favorable-autocomplete-results-report-shows-2016-06-09.

68   Daniel Trielli, Sean Mussenden and Nicholas Diakopoulos, "Why Google Search Results Favor Democrats," *Slate*, Dec. 7, 2015, http://www.slate.com/articles/technology/future_tense/2015/12/why_google_search_results_favor_democrats.html.

69   Edelman, "Hard-Coding Bias."

70   Brody Mullins, Rolfe Winkler and Brent Kendall, "Inside the U.S. Antitrust Probe of Google," *The Wall Street Journal*, March 19, 2015, https://www.wsj.com/articles/inside-the-u-s-antitrust-probe-of-google-1426793274.

71   "Left-Center Bias," *Media Bias/Fact Check*, accessed April 11, 2019, https://mediabiasfactcheck.com/leftcenter/.

72   Jayson Veley, "Confirmed: Google Punishing Conservative Websites With Search Engine Ranking Penalties," *Censorship News*, Sept. 22, 2017, http://www.censorship.news/2017-09-22-confirmed-google-punishing-conservative-websites-with-search-engine-ranking-penalties.html.

73   David Harsanyi, "Google's New 'Fact-Checker' Is Partisan Garbage," *The Federalist*, Jan. 10, 2018, http://thefederalist.com/2018/01/10/googles-new-factchecker-is-partisan-garbage/.

74   Monica Showalter, "Google Caught in the Censorship Cookie Jar Again?" *American Thinker*, Jan. 10, 2018, https://www.americanthinker.com/blog/2018/01/google_caught_in_the_censorship_cookie_jar_again_comments.html.

75   Phil McCausland, "ABC News Reporter Brian Ross Suspended for 'Serious Error' in Flynn Reporting," *NBCNews.com*, Dec. 2, 2017, https://www.nbcnews.com/news/us-news/abc-news-reporter-brian-ross-suspended-serious-error-flynn-reporting-n825966.

76   Charlie Nash, "Google Home Assistant Refuses to Define Jesus Christ, but No Problem With Muhammad, Buddha," *Breitbart*, Jan. 26, 2018, http://www.breitbart.com/tech/2018/01/26/video-google-home-assistant-refuses-define-jesus-christ-defines-muhammad-buddha/.

77   Kelen McBreen, "Google Home Censors Jesus Christ," *InfoWars*, Jan. 25, 2018, https://www.infowars.com/shock-video-google-home-censors-jesus-christ/.

78   Sites listed in this section are sourced from author interviews, *The Gateway Pundit*'s list of censored sites, and as otherwise footnoted.

79   Allum Bokhari, "Trump's Facebook Engagement Declined by 45 Percent Following Algorithm Change," *Breitbart*, Feb. 28, 2018, https://www.breitbart.com/tech/2018/02/28/exclusive-trumps-facebook-engagement-declined-45-percent-following-algorithm-change/.

80   Mike Adams, "Facebook Bans the Declaration of Independence, Flags It as 'Hate Speech," *Censorship News*, July 5, 2018, http://censorship.news/2018-07-05-facebook-bans-the-declaration-of-independence-flags-it-as-hate-speech.html.

81   Heyes, "Facebook Slammed."

82   Ian Miles Cheong, "Conservative and Independent YouTube Channels Hit by Censorship and Demonetization," *The Daily Caller*, Aug. 11, 2017, http://dailycaller.com/2017/08/11/conservative-and-independent-youtube-channels-hit-by-censorship-and-demonetization/.

83   Calvin Freiburger, "Baby Pics 'Too Strong for Facebook': Social Media Giant Targets Pro-Lifers in New Ad Rules," *Life Site*, June 28, 2018, https://www.lifesitenews.com/news/baby-pics-too-strong-for-facebook-social-media-giant-targets-pro-lifers-in.

84   Janko Roettgers, "Google Hires 4Chan Founder Chris Poole," *Variety*, March 7, 2016, https://variety.com/2016/digital/news/google-hires-4chan-founder-chris-poole-moot-1201724308/.

85   Cheong, "YouTube Channels."

86   JD Heyes, "True FASCISM: Here's the List of Conservative News Sites and Figures Who Are Being Shadow-Banned on Facebook, Google," *Censorship News*, March 5, 2018, http://www.censorship.news/2018-03-05-fascism-list-of-conservative-news-sites-shadow-banned-on-facebook-google.html.

87   Cheong, "YouTube Channels."

88   Ibid.

89   Leo Goldstein, "Google's Search Bias Against Conservative News Sites Has Been Quantified," *Watts Up With That?* Sept. 8, 2017, https://wattsupwiththat.com/2017/09/08/a-method-of-google-search-bias-quantification-and-its-application-in-climate-debate-and-general-political-discourse/.

90   "Google Preaches 'Net Neutrality,' Then Censors Conservative Videos?" *Investor's Business Daily*, Editorial, Oct. 26, 2017, https://www.investors.com/politics/editorials/google-should-practice-what-it-preaches-on-net-neutrality/.

91   "PragerU Takes Legal Action Against Google and YouTube for Discrimination," PragerU.com, news release, accessed April 14, 2019, https://www.prageru.com/press-release-prager-university-prageru-takes-legal-action-against-google-and-youtube-discrimination.

92   Asche Schow, "YouTube Censoring Videos From Conservative Group," *Washington Examiner*, Oct. 12, 2016, http://www.washingtonexaminer.com/youtube-censoring-videos-from-conservative-group/article/2604384.

93   "Google," *Investor's Business Daily*.

94   Eric Lieberman, "Judge Tosses PragerU Lawsuit Accusing Google, YouTube of Censoring Conservative Content," *The Daily Signal*, March 29, 2018, https://www.dailysignal.

com/2018/03/29/judge-tosses-prageru-lawsuit-accusing-google-youtube-of-censor-ing-conservative-content/.

95   "PragerU Doubles Down on Censorship Lawsuit Vs. Google/YouTube in State Court of California." PragerU, Jan. 8, 2019. https://www.prageru.com/press-release/prageru-doubles-down-on-censorship-lawsuit-vs-google-youtube-in-state-court/.

96   Upper and Hair, "Confirmed."

97   Terence Cullen and Janon Fisher, "Florida School Shooting Survivors Hit Lawmakers Who Oppose Gun Control With Ultimatum," *New York Daily News*, Feb. 18, 2018, http://www.nydailynews.com/news/national/students-plan-nationwide-gun-control-march-florida-shooting-article-1.3828101.

98   Chris Perez, "Florida School Shooting Suspect Is 'Troubled' Former Student Obsessed With Guns," *New York Post*, Feb. 14, 2018, https://nypost.com/2018/02/14/florida-school-shooting-suspect-is-troubled-former-student-obsessed-with-guns.

99   Meghann Farnsworth, "Watch the Interview: Facebook's Heads of News Feed and News Partnerships," *Recode*, Feb. 18, 2018, https://www.recode.net/2018/2/9/16996696/how-to-watch-livestream-facebook-head-of-news-feed-partnerships-campbell-brown-adam-mosseri.

100   Sara Fischer, "Rare, Cox Media's Facebook-Driven Conservative Site, Is Shutting Down," Axios, March 5, 2018, https://www.axios.com/cox-media-rare-will-shut-down-facebook-1520265673-029db27d-1c6d-4d32-b5ec-2ce47523c000.html.

101   Farnsworth, "Watch."

102   To conduct this evaluation, *The Western Journal* pulled Facebook data from Crowd-Tangle, which is owned by Facebook, for all members of this current Congress with a Facebook page. That data was aggregated for Facebook pages from August 2017 through June 2018. The analysis described herein does not include data from Facebook pages that did not post during any one of the eleven months of data pulled. Out of the 577 congressional Facebook pages available, eighty-one were eliminated from the analysis because of this. The two independent senators, Sens. Bernie Sanders and Angus King, were included with the Democrats because they caucus with the Democratic Party. *The Western Journal* then took the data from CrowdTangle and calculated each Congress member's monthly interaction rate using the total interactions on the page, total posts, and total page likes, with all three weighted equally. The total interactions are the total number of reactions, shares and comments on a Facebook post. The interaction rate was calculated by averaging the number of interactions for all of the account's posts in the specified time frame and then dividing that number by the number of followers of that page. The pre-algorithm change data includes all data from August through December 2017; the post-change data includes all data from February through June 2018. The data used for this analysis measures users' interactions with the posts and not the reach of the post. Reach data is available only to individual publishers and is not made public by Facebook. However, the interactions are good general indicators of reach because when more users see a given post, interactions with that post should rise accordingly.

103   George Upper and Erin Coates, "Study: US Congress Members Hurt After Facebook Algorithm Change, But Republicans Hit Harder," *Western Journal*, July 23, 2018, https://www.westernjournal.com/congress-facebook-algorithm-study-2018/.

104 Total video interactions in this case does not include Facebook Live videos because those videos were not yet being used in any meaningful number by members of Congress.

105 Charles Duhigg, "The Case Against Google," *The New York Times*, February 20, 2018, https://www.nytimes.com/2018/02/20/magazine/the-case-against-google.html.

106 Ibid.

107 To learn more about this talented writer, visit his homepage at https://charlesduhigg.com/about/ and find his columns regularly in *The New York Times*.

108 Ibid.

109 *Tucker Carlson Tonight*, "Social Media Network CEO: How Google Censors My Company," Fox News, Feb. 21, 2018, http://video.foxnews.com/v/5738232786001/?#sp=show-clips.

110 List of Mergers and Acquisitions by Alphabet. In Wikipedia, accessed April 14, 2019, https://en.wikipedia.org/wiki/List_of_mergers_and_acquisitions_by_Alphabet.

111 Robert Epstein, "The Unprecedented Power of Digital Platforms to Control Opinions and Votes," Pro Market (blog of the Stigler Center at the University of Chicago Booth School of Business), April 12, 2018, https://promarket.org/unprecedented-power-digital-platforms-control-opinions-votes/.

112 Nitasha Tiku and Casey Newton, "Twitter CEO: 'We Suck at Dealing With Abuse,'" *The Verge*, Feb. 4, 2015, https://www.theverge.com/2015/2/4/7982099/twitter-ceo-sent-memo-taking-personal-responsibility-for-the.

113 Sean J. Edgett, testimony before the United States Senate Committee on the Judiciary, Subcommittee on Crime and Terrorism, Oct. 31, 2017, https://www.lgraham.senate.gov/public/_cache/files/4766f54d-d433-4055-9f3d-c94f97eeb1c0/testimony-of-se-an-edgett-acting-general-counsel-twitter.pdf.

114 Nick Douglas, "What Is Shadow Banning on Twitter?" *Lifehacker*, July 30, 2018, https://lifehacker.com/did-twitter-shadow-ban-you-1827972917.

115 "Twitter Engineers To 'Ban a Way of Talking' Through 'Shadow Banning,' Algorithms to Censor Opposing Political Opinions," Project Veritas, Jan. 11, 2018, https://www.projectveritas.com/2018/01/11/undercover-video-twitter-engineers-to-ban-a-way-of-talking-through-shadow-banning-algorithms-to-censor-opposing-political-opinions/.

116 Ibid.

117 Milo Yiannopoulos, "Twitter Shadowbanning 'Real and Happening Every Day' Says Inside Source," *Breitbart*, Feb. 16, 2016, https://www.breitbart.com/tech/2016/02/16/exclusive-twitter-shadowbanning-is-real-say-inside-sources/.

118 James Freeman, "Can We Trust Facebook and Twitter?" *The Wall Street Journal*, July 31, 2018, https://www.wsj.com/articles/can-we-trust-facebook-and-twitter-1533074266.

119 Brett Samuels, "Trump: 'We Will Look Into' Twitter for 'Shadow Banning' Republicans," *The Hill*, July 26, 2018, http://thehill.com/policy/technology/398943-trump-government-will-look-into-twitter-for-shadow-banning-republicans.

120 Adam Candeub, "A Response to Online Shadow Banning," *The Wall Street Journal*, Aug. 5, 2018, https://www.wsj.com/articles/a-response-to-online-shadow-banning-1533496357.

121 Doug Mainwaring, "'Shadow Banning': How Twitter Secretly Censors Conservatives Without Them Even Knowing It," Life Site News, January 12, 2018, https://www.lifesitenews.com/news/shadow-banning-how-twitter-secretly-censors-conservatives-without-them-even.

122 Bradford Richardson, "Twitter Suppresses Messages Criticizing Planned Parenthood, Pro-Life Group Says," *The Washington Times*, June 27, 2017, https://www.washington-times.com/news/2017/jun/27/twitter-suppresses-messages-criticizing-planned-pa/.

123 *Tucker Carlson Tonight*, "Founder of Pro-Life Group Blasts Twitter CEO," Fox News, Aug. 31 2018, http://insider.foxnews.com/2018/09/01/twitter-accused-shadow-ban-ning-conservative-pro-life-group-founder-blasts-ceo-jack-dorsey.

124 Allysia Finley, "OK Google, You've Been Served," *The Wall Street Journal*, Jan. 15, 2018, https://www.wsj.com/articles/ok-google-youve-been-served-1516044100.

125 California Labor Code, Division 2, Part 3, Chapter 5 (Political Affiliations), accessed April 14, 2019, https://leginfo.legislature.ca.gov/faces/codes_displaySection.xhtml?lawCode=LAB&sectionNum=1101.

126 Finley, "OK Google."

127 Allum Bokhari, "Rebels of Google: Senior Management 'On the Verge of Tears' After Trump Win," *Breitbart*, Aug. 9, 2017, https://www.breitbart.com/tech/2017/08/09/rebels-of-google-management-verge-of-tears-trump-win/.

128 Finley, "OK Google."

129 James Damore et al. v. Google, class action complaint, Superior Court of California, accessed April 14, 2019, https://www.dhillonlaw.com/wp-content/uploads/2018/04/20180418-Damore-et-al.-v.-Google-FAC_Endorsed.pdf.

130 Bokhari, "Rebels of Google."

131 Erin Griffith, "'Sex Party' or 'Nerds on a Couch?' A Night in Silicon Valley," *Wired*, Jan. 11, 2018, https://www.wired.com/story/sex-party-or-nerds-on-a-couch-a-night-in-silicon-valley.

132 Emily Chang, "'Oh My God, This Is So F---ed Up': Inside Silicon Valley's Secretive, Orgiastic Dark Side," *Vanity Fair*, February 2018, https://www.vanityfair.com/news/2018/01/brotopia-silicon-valley-secretive-orgiastic-inner-sanctum.

133 Sara Ashley O'Brien, "Google Hit With Revised Gender Pay Lawsuit," *CNN.com*, Jan. 3, 2018, https://money.cnn.com/2018/01/03/technology/google-gender-pay-lawsuit-revised/index.html.

134 Dean Takahashi, "How Publishers and Advertisers Can Balance Privacy and Monetization," *Venture Beat*, July 14, 2018, https://venturebeat.com/2018/07/14/how-publishers-and-advertisers-can-balance-privacy-and-monetization/.

135 "If You Had Invested Right After Apple's IPO," *Investopedia*, Updated Nov. 2, 2018, https://www.investopedia.com/articles/active-trading/080715/if-you-would-have-invested-right-after-apples-ipo.asp.

136 Megan Leonhardt, "If You Invested $1,000 in Amazon in 1997, Here's How Much You'd Have Now," *CNBC.com*, Aug. 31, 2018, https://www.cnbc.com/2018/08/31/if-you-put-1000-dollars-in-amazon-in-1997-heres-how-much-youd-have-now.html.

137 Jeff Desjardins, "The Largest Companies by Market Cap Over 15 Years," *Visual Capitalist*, Aug. 12, 2016, http://www.visualcapitalist.com/chart-largest-companies-market-cap-15-years/.

138 David Rotman, "Technology and Inequality," *MIT Technology Review*, Oct. 21, 2014, https://www.technologyreview.com/s/531726/technology-and-inequality.

139 Jason Douglas, Jon Sindreu and Georgi Kantchev, "The Problem With Innovation: The Biggest Companies Are Hogging All the Gains," *The Wall Street Journal*, July 15, 2018, https://www.wsj.com/articles/the-problem-with-innovation-the-biggest-companies-are-hogging-all-the-gains-1531680310.

140 Ibid.

141 Vauhini Vara, "Can This Startup Break Big Tech's Hold on A.I.?" *Fortune*, June 25, 2018, http://fortune.com/longform/element-ai-startup-tech/.

142 Olivia Solon, "When Should a Tech Company Refuse to Build Tools for the Government?" *The Guardian*, June 26, 2018, https://www.theguardian.com/technology/2018/jun/26/tech-government-contracts-worker-revolt-microsoft-amazon-google.

143 Farai Chideya, "Nearly All of Silicon Valley's Political Dollars Are Going to Hillary Clinton," *FiveThirtyEight*, Oct. 25, 2016, https://fivethirtyeight.com/features/nearly-all-of-silicon-valleys-political-dollars-are-going-to-hillary-clinton/.

144 "An Open Letter From Technology Sector Leaders on Donald Trump's Candidacy for President," published at *NewCo Shift*, July 14, 2016, https://shift.newco.co/an-open-letter-from-technology-sector-leaders-on-donald-trumps-candidacy-for-president-5bf734c159e4.

145 Kate Crawford, "Artificial Intelligence's White Guy Problem," *The New York Times*, June 26, 2016, http://www.nytimes.com/2016/06/26/opinion/sunday/artificial-intelligences-white-guy-problem.html.

146 Ibid.

147 Rob Leathern, "Updates to Our Prohibited Financial Products and Services Policy," on Facebook, June 26, 2018, https://www.facebook.com/business/news/updates-to-our-prohibited-financial-products-and-services-policy.

148 Mark Epstein, "Google's and Facebook's Dubious Bitcoin Bans," *The Wall Street Journal*, June 25, 2018, https://www.wsj.com/articles/googles-and-facebooks-dubious-bitcoin-bans-1529966342.

149 Ibid.

150 Ibid.

151 Tunku Varadarajan, "Sage Against the Machine," *The Wall Street Journal*, Aug. 31, 2018, https://www.wsj.com/articles/sage-against-the-machine-1535747443.

OUT TO THE WOODSHED

1 Jeffrey Gottfried and Elisa Shearer, "Americans' Online News Use Is Closing In on TV News Use," Pew Research Center, Sept. 7, 2017, http://www.pewresearch.org/fact-tank/2017/09/07/americans-online-news-use-vs-tv-news-use; Greg Sterling, "Google Controls 65 Percent Of Search, Bing 33 Percent," *Search Engine Land*, Aug. 21, 2015, https://searchengineland.com/google-controls-65-percent-of-search-bing-33-percent-comscore-228765.

2 Martin Giles, "It's Time to Rein In the Data Barons," *MIT Technology Review*, June 19, 2018, https://www.technologyreview.com/s/611425/its-time-to-rein-in-the-data-barons.

3 Ip, "Antitrust Case."

4 Ibid.

5 "Apple Vs Android–A Comparative Study 2017," Moon Technolabs, accessed April 14, 2019, https://www.moontechnolabs.com/apple-vs-android-comparative-study-2017/.

6 Ip, "Antitrust Case."

7 Jake Kanter, "Meet the Campaigners Who Say They Will Stop at Nothing in Their Quest to Break Up Facebook," *Yahoo*, July 22, 2018, https://www.yahoo.com/amphtml/finance/news/meet-campaigners-stop-nothing-quest-070000380.html.

8 Morris, "The New Tech Avengers."

9 Kerry, "Protecting Privacy."

10   Rana Foroohar, "Google's Abuse of Power Reveals the Modern Disease of Oligopolies," *Financial Review*, July 23, 2018, https://www.afr.com/technology/apps/business/googles-abuse-of-power-reveals-the-modern-disease-of-oligopolies-20180722-h130d8.

11   Ibid.

12   Mark Warner, "Potential Policy Proposals for Regulation of Social Media and Technology Firms," *Axios*, accessed April 14, 2019, https://graphics.axios.com/pdf/PlatformPolicyPaper.pdf.

13   David McCabe, "Scoop: 20 Ways Democrats Could Crack Down on Big Tech," *Axios*, July 30, 2018, https://www.axios.com/mark-warner-google-facebook-regulation-policy-paper-023d4a52-2b25-4e44-a87c-945e73c637fa.html.

14   Tony Romm, "The Trump Administration Is Talking to Facebook and Google About Potential Rules for Online Privacy," *The Washington Post*, July 27, 2018, https://www.washingtonpost.com/technology/2018/07/27/trump-administration-is-working-new-proposal-protect-online-privacy.

15   Roberts, "Manhattan Store Owner"; David McLaughlin, "Amazon Antitrust Critic Joins FTC as Agency Sets Sights on Tech," *Bloomberg*, July 9, 2018, https://www.bloomberg.com/news/articles/2018-07-09/amazon-antitrust-critic-joins-ftc-as-agency-sets-sights-on-tech.

16   Sam Schechner and Nick Kostov, "Facebook's Next Privacy Challenge: Less Data to Target Ads," *The Wall Street Journal*, July 30, 2018, https://www.wsj.com/articles/facebooks-next-privacy-challenge-less-data-to-target-ads-1532943001.

17   Christopher Mims, "Tech's Titans Tiptoe Toward Monopoly," *The Wall Street Journal*, May 31, 2018, https://www.wsj.com/articles/techs-titans-tiptoe-toward-monopoly-1527783845.

18   Randall E. Stross, "How Yahoo! Won The Search Wars," *Fortune*, March 2, 1998, http://archive.fortune.com/magazines/fortune/fortune_archive/1998/03/02/238576/index.htm.

19   William Rinehart, "Breaking Up Big Tech Is Hard to Do," *The Wall Street Journal*, July 22, 2018, https://www.wsj.com/articles/breaking-up-big-tech-is-hard-to-do-1532290123.

20   Scott Galloway, "Bread Crumbs," *L2Inc.com*, Sept. 8, 2017, https://www.l2inc.com/daily-insights/no-mercy-no-malice/bread-crumbs.

21   Margaret Visnji, "Amazon vs Walmart Revenues and Profits 1995-2014," *Revenues and Profits*, Jan. 22, 2019, https://revenuesandprofits.com/amazon-vs-walmart-revenues-and-profits-1995-2014.

22   Manuela Tobias, "No, the Postal Service Isn't Losing a Fortune on Amazon," *Politifact*, April 2, 2018, https://www.politifact.com/truth-o-meter/statements/2018/apr/02/donald-trump/trump-usps-postal-service-amazon-losing-fortune/.

23   Emily Stewart, "Happy Prime Day! Experts worry Amazon is building a dangerous monopoly," Vox, July 17, 2018, https://www.vox.com/2018/7/17/17583070/amazon-prime-day-monopoly-antitrust.

24   Ibid.

25   Ibid.

26   Terrence Dopp, Justin Sink, and Ben Brody, "Trump Says Bezos Guards Amazon on Antitrust With Washington Post," *Bloomberg*, July 23, 2018, https://www.bloomberg.com/news/articles/2018-07-23/trump-accuses-washington-post-of-being-lobbyist-for-amazon.

27   "Don't Break Up Facebook," *Bloomberg*, Editorial, July 16, 2018, https://www. bloomberg.com/view/articles/2018-07-16/don-t-break-up-facebook.

28   "The Techlash Against Amazon, Facebook and Google—and What They Can Do," *The Economist*, Jan. 20, 2018, https://www.economist.com/briefing/2018/01/20/ the-techlash-against-amazon-facebook-and-google-and-what-they-can-do.

29   Scott Cleland, "Break Up Google: There Is a Solid Conservative Antitrust Case Against Alphabet-Google," *The Daily Caller*, Jan. 19, 2018, http://dailycaller.com/2018/01/19/ break-up-google-there-is-a-solid-conservative-antitrust-case-against-alphabet-google/.

30   Ibid.

31   Ibid.

32   Ricardo Soltero, "Walmart's New Tech Will Be Able to Spy on Employees and Customers," *iDrop News*, July 16, 2018, https://www.idropnews.com/news/fast-tech/ walmarts-new-tech-will-be-able-to-spy-on-employees-and-customers/76599/.

33   "Should the U.S. Follow Europe in Imposing Stricter Data-Privacy Regulations?" *The Wall Street Journal*, June 18, 2018, https://www.wsj.com/articles/should-the-u-s-follow-europe-in-imposing-stricter-data-privacy-regulations-1529332491

34   Justin Caruso, "Peter Thiel Says Silicon Valley 'a Totalitarian Place'—Slams 'Political Correctness'," *The Daily Caller*, March 16, 2018,
https://dailycaller.com/2018/03/16/peter-thiel-silicon-valley-political-correctness/.

35   Kerry, "Protecting Privacy."

36   Duhigg, "Case Against Google."

About the Authors

FLOYD BROWN is a business innovator, writer, and speaker. He currently serves as the publisher of WesternJournal.com, one of America's largest online news organizations.

Brown is the author of six books and has written for many publications, including the *San Francisco Chronicle*, the *Washington Times*, Townhall.com, and *National Review*.

Brown has appeared on many TV networks and shows including: *The O'Reilly Factor*, the CBS Evening News, ABC's *Primetime*, NBC's *Today*, Fox News, CNN, MSNBC, and more.

TODD CEFARATTI has been a successful entrepreneur, CEO, and conservative political activist in Washington D.C. for the past thirty years. Cefaratti is a graduate of UCLA where he received his Bachelor Arts degree in Economics and Business and he also attended Harvard Business School where he studied digital marketing. Cefaratti is a Senior Director/Consultant at one of the largest national conservative news organizations in the country with an audience over thirty million, Liftable Media. Cefaratti is also the Co-Founder of a new free speech news and social platform FeedMe (feedme.app).

Cefaratti has also appeared on many TV and radio networks and shows, including: Fox News, *For the Record with Greta Van Susteren*, RT TV, *The Blaze*, *The Rusty Humphries Show*, *The Washington Times*, Washington Times Radio, TPNN, Newsmax.com, and many more.